U0161476

向阳而生

随心手账

[英] 海伦·埃克斯利——编选
周成刚——主编

Every day is a new day.

每一天，都是新的开始。

Mon. / Tue. / Wed. / Thur. / Fri. / Sat. / Sun.

Date

年

月

日

In character, in manner, in style, in all things, the supreme excellence is simplicity.

性格、态度、风格以及其他一切事物，简单的就是最好的。

——亨利·沃兹沃斯·朗费罗

（HENRY WADSWORTH LONGFELLOW）

Mon. / Tue. / Wed. / Thur. / Fri. / Sat. / Sun.

Date

年

月

日

How simple and frugal a thing is happiness: a glass of wine, a roast chestnut, a wretched little brazier, the sound of the sea... All that is required to feel that here and now is happiness is a simple, frugal heart.

幸福是多么简单质朴：一杯红酒、一颗烤栗子、一个破旧的小火盆、大海的声音……要想感受到此时此地的幸福，只需要一颗简单而质朴的心。

——尼可斯·卡赞扎基斯（NIKOS KAZANTZAKIS）

Mon. / Tue. / Wed. / Thur. / Fri. / Sat. / Sun.

Date

年

月

日

Contentment consists not in adding more fuel, but in taking away some fire; not in multiplying of wealth, but in subtracting our desires.

知足不在于添柴加薪，而在于不让火烧得过旺；它也不在于使财富倍增，而在于减少欲望。

——托马斯·富勒

（THOMAS FULLER）

Mon. / Tue. / Wed. / Thur. / Fri. / Sat. / Sun.

Date

年

月

日

Simplicity is making the journey of this life with just baggage enough.

简单，就是在人生的旅途中，只带着足够用的行李。

——查尔斯·杜德利·华纳

（CHARLES DUDLEY WARNER）

Mon. / Tue. / Wed. / Thur. / Fri. / Sat. / Sun.

Date

年

月

日

When we are truly in this interior simplicity our whole appearance is franker, more natural. This true simplicity makes us conscious of a certain openness, gentleness, innocence, gaiety and serenity.

当我们真正拥有内心的简单，我们的外表就会变得更率真自然。这种真正的简单让我们感受到一种开阔、温柔、天真、快乐和宁静。

——弗朗索瓦 · 费奈隆

（FRANCOIS FENELON）

Mon. / Tue. / Wed. / Thur. / Fri. / Sat. / Sun.

Date

年

月

日

A small house will hold as much happiness as a big one.

小屋子容纳的幸福与大房子一样多。

佚名

（AUTHOR UNKNOWN）

Date

年

月

日

Live in each season as it passes, breathe the air, drink the drink, taste the fruit, and resign yourself to the influences of each.

活于当下，品味四季。呼吸新鲜的空气，畅饮杯中的美酒，品尝当季的水果，尽情享受这一切的美妙吧。

——亨利·戴维·梭罗

（HENRY DAVID THOREAU）

Mon. / Tue. / Wed. / Thur. / Fri. / Sat. / Sun.

Date

年

月

日

The miracle is not to fly in the air, or to walk on the water, but to walk on the earth.

奇迹不是在稀薄的空气中或水面上行走，而是在大地上行走。

——谚语

（PROVERB）

Mon. / Tue. / Wed. / Thur. / Fri. / Sat. / Sun.

Date

年

月

日

When your schedule leaves you brain-drained and stressed to exhaustion, it's time to give up something. Delegate. Say no. Be brutal. It's like cleaning out a closet – after a while, it gets easier to get rid of things. You discover that you really didn't need them anyway.

当你的日程安排让你思绪枯竭、精疲力尽时，就应该放弃一些东西了。分派任务，敢于拒绝，处理事情直截了当。就像清理壁橱一样——不一会儿，扔掉东西就容易多了。你会发现，其实你根本不需要它们。

——玛丽莲·鲁曼（MARILYN RUMAN）

Mon. / Tue. / Wed. / Thur. / Fri. / Sat. / Sun.

Date

年

月

日

I have found such joy in simple things: a plain, clean room, a nut-brown loaf of bread, a cup of milk, a kettle as it sings, the shelter of a roof above my head, and in a leaf-laced square along the floor, where yellow sunlight glimmers through a door.

在一些简单的事物中，我找到了无尽的快乐：一个朴素干净的房间、一条栗色面包、一杯牛奶、一只吱吱叫的水壶、一处遮风挡雨的屋顶，还有一块镶着树叶的地砖。在那里，金色的阳光透过门缝照射进来。

——格雷丝·诺尔·克罗韦尔

（GRACE NOLL CROWELL）

Date

年

月

日

The secret of contentment is knowing how to enjoy what you have, and to be able to lose all desire for things beyond your reach.

知足常乐的秘诀是：懂得如何享用你所拥有的，并割舍不实际的欲念。

——林语堂

（LIN YUTANG）

Mon. / Tue. / Wed. / Thur. / Fri. / Sat. / Sun.

Date

年

月

日

I want to work. At what? I want to live that I may work with my hands and my feeling and my brain. I want a garden, a small house, grass, animals, books, pictures, music.

我想工作。做什么呢？我想要过这样的生活：我可以用我的手、我的感觉和我的大脑工作。我想要一个花园、一所小房子，有植物、动物、书籍、图画和音乐。

——凯瑟琳 · 曼斯菲尔德

（KATHERINE MANSFIELD）

Mon. / Tue. / Wed. / Thur. / Fri. / Sat. / Sun.

Date

年

月

日

…from the sky, from the earth, from a scrap of paper, from a passing shape, from a spider's web…We must pick out what is good for us where we can find it.

在天上，在地上，在纸片上，在匆匆闪过的影子上，在蜘蛛网上……如果有可能，我们必须在各样事物中找出对我们有益的东西。

——巴勃罗·毕加索

（PABLO PICASSO）

Mon. / Tue. / Wed. / Thur. / Fri. / Sat. / Sun.

Date

年

月

日

Before enlightenment, chop wood, carry water; After enlightenment, chop wood, carry water.

开悟之前，砍柴挑水；开悟之后，砍柴挑水。

——禅宗名言

（ZEN SAYING）

Date

年

月

日

It is not how much we have, but how much we enjoy, that makes happiness.

幸福，不在于拥有多少，而在于享受多少。

——查尔斯·哈登·司布真

（CHARLES H. SPURGEON）

Mon. / Tue. / Wed. / Thur. / Fri. / Sat. / Sun.

Date

年

月

日

As for me if the world turns this way I take it; if it turns another way I take it. Any way the world turns I take it with my hands.

对我而言，如果世界变成一种样子，我欣然接受；如果世界变成另一种样子，我也欣然接受。无论世界怎样转变，我都会张开双手去拥抱它。

——加布里埃尔·奥卡拉

（GABRIEL OKARA）

Mon. / Tue. / Wed. / Thur. / Fri. / Sat. / Sun.

Date

年

月

日

The nicest and sweetest days are not those on which anything very splendid or wonderful or exciting happens, but just those that bring simple little pleasures, following one another softly, like pearls slipping off a string.

最甜蜜美好的日子，不是那些精彩非凡或激动人心的日子，而是那些平凡却又幸福的日子。这些日子一个接着一个，就像珍珠从细线中轻柔地滑落。

——露西·莫德·蒙哥马利

（LUCY MAUD MONTGOMERY）

Mon. / Tue. / Wed. / Thur. / Fri. / Sat. / Sun.

Date

年

月

日

Anyone who piles up treasure has much to lose.

多藏必厚亡。

——老子

(LAO TZU)

Date

年

月

日

The simple things are often best: a simple, helping hand, a kindly thought, the simple phrase, "Of course, I understand."

简单的事情往往是最美好的：一双援助之手，一个善良的念头，一句简单的话语——"当然，我懂的。"

——安妮·克雷尔

（ANNE KREER）

Mon. / Tue. / Wed. / Thur. / Fri. / Sat. / Sun.

Date

年

月

日

I loafe and invite my soul...

我邀了我的灵魂同我一道闲游……

——沃尔特·惠特曼

（WALT WHITMAN）

Date

年

月

日

The best person is like water. Water is good; it benefits all things and does not compete with them. It dwells in lowly places that all disdain.

上善若水。水善利万物而不争，处众人所恶，故几于道。

——老子

（LAO TZU）

Mon. / Tue. / Wed. / Thur. / Fri. / Sat. / Sun.

Date

年

月

日

Life has an elegance that far exceeds anything we might devise. Perhaps the wisdom lies in knowing when to sit back and wait for it to unfold.

生活的优雅美妙远远超出我们的想象。也许智慧的真谛在于，明白该在什么时候停下来，等待生活慢慢绽放。

——蕾切尔·娜奥米·雷蒙

（RACHEL NAOMI REMEN）

Mon. / Tue. / Wed. / Thur. / Fri. / Sat. / Sun.

Date

年

月

日

The only difference between an extraordinary life and an ordinary one is the extraordinary pleasures you find in ordinary things.

非凡人生和平凡人生之间唯一的区别是：是否能从平凡的事物中找到不平凡的快乐。

——薇若妮卡·魏纳

（VERONIQUE VIENNE）

Mon. / Tue. / Wed. / Thur. / Fri. / Sat. / Sun.

Date

年

月

日

Much that I sought, I could not find; much that I found, I could not bind; much that I bound, I could not free; much that I freed, returned to me.

当我追求很多时，我无法寻得；当我寻得很多时，我无法让它们结合；当我把很多东西结合到一起时，我无法舍弃它们；当我舍弃很多时，我才回归了自我。

——李·威尔逊·多德
（LEE WILSON DODD）

Mon. / Tue. / Wed. / Thur. / Fri. / Sat. / Sun.

Date

年

月

日

To have peace in one's soul is the greatest happiness.

极致的快乐源于内心的宁静。

——东方箴言（ORIENTAL WISDOM）

Mon. / Tue. / Wed. / Thur. / Fri. / Sat. / Sun.

Date

年

月

日

When the sun rises, I go to work. When the sun goes down, I take my rest. I dig the well from which I drink, I farm the soil which yields my food, I share creation; kings can do no more.

日出而作，日入而息。凿井而饮，耕田而食。帝力于我何有哉！

——谚语

（PROVERB）

Mon. / Tue. / Wed. / Thur. / Fri. / Sat. / Sun.

Date

年

月

日

We do not want riches. We want peace and love.

我们不想要财富，我们要的是爱与宁静。

——红云

（RED CLOUD）

Mon. / Tue. / Wed. / Thur. / Fri. / Sat. / Sun.

Date

年

月

日

Simplify. Stop bothering with non-essentials. Having devoted my life to my work so far, I should reap the harvest and learn how to live the rest of it properly. You can be lazy at last, so enjoy yourself, man. It's time now for trees and grass and growing things.

变得简单些。不再为无关紧要的事情伤神。一直以来，我都全身心地投入工作，现在到了收获的季节，我应该学习如何恰当地度过余生了。终于可以偷懒了，尽情享受吧！是时候莳花弄草了。

——佚名（AUTHOR UNKNOWN）

Mon. / Tue. / Wed. / Thur. / Fri. / Sat. / Sun.

Date

年

月

日

To live content with small means; to seek elegance rather than luxury, and refinement rather than fashion; to be worthy, not respectable, and wealthy, not rich; to study hard, think quietly, talk gently, act frankly...

收入不高，但要活得知足；追求优雅而不是奢华；追求教养，而非时尚；要有价值，但不必被社会认可；要富足，但不用特别富有；努力学习，安静思考，温柔交谈，坦诚待人……

——威廉·埃勒里·钱宁

（WILLIAM ELLERY CHANNING）

Mon. / Tue. / Wed. / Thur. / Fri. / Sat. / Sun.

Date

年

月

日

A happy life is made up less of great events than little, lovely moments.

幸福的生活不是由重大事件组成的，而是由各种美好的微小瞬间组成的。

——帕姆·布朗

（PAM BROWN）

Mon. / Tue. / Wed. / Thur. / Fri. / Sat. / Sun.

Date

年

月

日

May we never let the things we can't have, or don't have, or shouldn't have, spoil our enjoyment of the things we do have and can have.

希望我们永远不要让不曾拥有、不能拥有、不该拥有的东西，影响我们享受所拥有且能够拥有的东西。

——理查德·路易斯·埃文斯

（RICHARD L. EVANS）

Mon. / Tue. / Wed. / Thur. / Fri. / Sat. / Sun.

Date

年

月

日

Too much and we become confused and lose the ability to see the perfection of one simple thing.

东西太多，我们就会陷入困惑，看不到一件简单事物的完美了。

——帕姆·布朗

（PAM BROWN）

Date

年

月

日

Money is for buying the fruits of the earth, of the land where you were born.

金钱应当用来购买你生于斯长于斯的这片土地上的果实。

——艾伦·斯图尔特·佩顿

（ALAN STEWART PATON）

Mon. / Tue. / Wed. / Thur. / Fri. / Sat. / Sun.

Date

年

月

日

Sooner or later we all discover that the important moments in life are not... the birthdays, the graduations, the weddings, not the great goals achieved. The real milestones... come to the door of memory unannounced, stray dogs that amble in, sniff around a bit, and simply never leave.

或迟或早，我们都会发现，生命中最重要的时刻并不是……生日、毕业典礼、婚礼，也不是实现伟大目标时。真正的里程碑……往往不打一声招呼就闯入记忆之门，像流浪狗一样慢慢悠悠地进来，四处嗅嗅，就再也不离开了。

——苏珊·B. 安东尼（SUSAN B. ANTHONY）

Mon. / Tue. / Wed. / Thur. / Fri. / Sat. / Sun.

Date

年

月

日

Simple things such as health, freedom and a home where we are loved are priceless.

简单的东西是无价的，比如健康、自由，和一个令我们感受到爱的家。

——佚名

（AUTHOR UNKNOWN）

Mon. / Tue. / Wed. / Thur. / Fri. / Sat. / Sun.

Date

年

月

日

Reduce the complexity of life by eliminating the needless wants of life, and the labors of life reduce themselves.

消除不必要的欲望后，生活就不那么复杂了，生命的负重也就减轻了。

——艾温·威·蒂尔

（EDWIN WAY TEALE）

Mon. / Tue. / Wed. / Thur. / Fri. / Sat. / Sun.

Date

年

月

日

If you don't enjoy what you have, how could you be happier with more?

如果你连自己现在所拥有的都无法享受，你怎么会因为拥有更多而快乐呢？

——佚名

（AUTHOR UNKNOWN）

Mon. / Tue. / Wed. / Thur. / Fri. / Sat. / Sun.

Date

年

月

日

Besides the noble art of getting things done, there is a nobler art of leaving things undone.

除了把事情做好的崇高艺术，还有把事情搁下不做的崇高艺术。

——林语堂

（LIN YUTANG）

Date

年

月

日

Simplicity, clarity, singleness: these are the attributes that give our lives power and vividness and joy.

简单、清晰、专一，这些都是给予生命力量、色彩和欢乐的特质。

——理查德·霍洛威

（RICHARD HALLOWAY）

Mon. / Tue. / Wed. / Thur. / Fri. / Sat. / Sun.

Date

年

月

日

I will arise and go now, and go to Innisfree, and a small cabin build there, of clay and wattles made: Nine bean-rows will I have there, a hive for the honeybee, and live alone in the bee-loud glade. And I shall have some peace there, for peace comes dropping slow, dropping from the veils of the morning...

我就要起身走了，到茵尼斯弗利岛，造座小茅屋在那里，枝条编墙糊上泥；我要养上一箱蜜蜂，种上九行豆角，独住在蜂声嗡嗡的林间草地。那儿安宁会降临我，安宁慢慢儿滴下来，从晨的面纱滴落……

——威廉·巴特勒·叶芝

（WILLIAM BUTLER YEATS）

Date

年

月

日

Small is beautiful.

小即是美。

——厄恩斯特·弗里德里希·舒马赫

（E. F. SCHUMACHER）

Mon. / Tue. / Wed. / Thur. / Fri. / Sat. / Sun.

Date

年

月

日

In my hut this spring there is nothing – there is everything.

这个春天，我的小屋里什么也没有，却什么都有。

——山口素堂

（SODO）

Mon. / Tue. / Wed. / Thur. / Fri. / Sat. / Sun.

Date

年

月

日

Einstein's three rules of work: 1) Out of clutter find simplicity. 2) From discord make harmony. 3) In the middle of difficulty lies opportunity.

爱因斯坦工作三原则：（1）从混乱中寻找简单；（2）从无序中探求和谐；（3）在困难中发现机会。

——阿尔伯特·爱因斯坦

（ALBERT EINSTEIN）

Mon. / Tue. / Wed. / Thur. / Fri. / Sat. / Sun.

Date

年

月

日

The world has sent us a thousand exotic tastes. The supermarket shelves are stacked with astonishments which at last overwhelm us – and make us long for past simplicities. Buttered toast. Boiled eggs. And an apple from the garden. The pleasures of childhood.

世界将千百种异域的风味送到了我们面前。超市的货架上堆满了让我们惊讶不已，并最终会不知不觉买下的东西——这令我们不由得向往起过去的简单生活。黄油面包，水煮鸡蛋，从花园里摘下的苹果。那童年的种种快乐。

——帕姆·布朗（PAM BROWN）

Date

年

月

日

A simple dinner in a poor person's house, without tapestries and purple, has smoothed the wrinkles from the anxious brow.

在穷人家里吃顿简单的晚餐，没有挂毯和紫袍，却能抚平焦虑时皱起的眉头。

——昆图斯·贺拉斯·弗拉库斯（HORACE）

Mon. / Tue. / Wed. / Thur. / Fri. / Sat. / Sun.

Date

年

月

日

…there is joy for me, as ever, in not moving at all, but just basking in the sun at the allotment, watching the butterflies start their summer dance, between the buddleia and the ornamental grasses.

对我来说，像往常一样，静静待着，在空地上晒晒太阳，看着蝴蝶在观赏植物间翩翩起舞，就是一种快乐。

——巴尼·巴德斯利

（BARNEY BARDSLEY）

Mon. / Tue. / Wed. / Thur. / Fri. / Sat. / Sun.

Date

年

月

日

Success makes life easier. It doesn't make living easier…

成功使生存更容易，却没有使生活更舒适。

——布鲁斯·斯普林斯汀

（BRUCE SPRINGSTEEN）

Mon. / Tue. / Wed. / Thur. / Fri. / Sat. / Sun.

Date

年

月

日

Some wonder how people can be so happy when they have nothing. But... I saw that what one has or doesn't have is an entirely relative concept. The joy of having my hair washed, the taste of a sweet cherry tomato – that is not "nothing".

有些人会想，一无所有的人怎么会如此快乐。但是……我发现"拥有"或"没有"是一种完全相对的概念。洗完头的清爽，小番茄甜甜的味道——那些可不能算是"没有"。

——戈尔迪·霍恩（GOLDIE HAWN）

Date

年

月

日

They make their pride in making their dinner cost much; I make my pride in making my dinner cost little.

他们以花很多钱做晚餐为傲，而我以尽量少花钱而自豪。

——亨利·戴维·梭罗

（HENRY DAVID THOREAU）

Mon. / Tue. / Wed. / Thur. / Fri. / Sat. / Sun.

Date

年

月

日

The tipi is much better to live in; always clean, warm in winter, cool in summer; easy to move. The "modern" man builds big house, cost much money, like big cage, shut out sun, can never move; always sick.

帐篷更适宜居住——永远干净整洁，冬暖夏凉，便于移动。"现代"人花很多钱去盖大房子。那房子就像大笼子，挡住外面的阳光，不能移动，住在里面的人还老是生病。

——飞鹰

（CHIEF FLYING HAWK）

Mon. / Tue. / Wed. / Thur. / Fri. / Sat. / Sun.

Date

年

月

日

The usefulness of a pot comes from its emptiness.

埏埴以为器，当其无，有器之用。

——老子

（LAO TZU）

Mon. / Tue. / Wed. / Thur. / Fri. / Sat. / Sun.

Date

年

月

日

The best place to succeed is where you are with what you have.

通往成功的最佳状态是，立足于你现在的位置，并用好自己的资源。

——查尔斯·迈克尔·施瓦布

（CHARLES M. SCHWAB）

Mon. / Tue. / Wed. / Thur. / Fri. / Sat. / Sun.

Date

年

月

日

My roses are my jewels, the sun and moon my clocks, fruit and water my food and drink.

玫瑰是我的珠宝，太阳和月亮是我的钟表，水果和清水是我的食物和饮料。

——海斯特·露西·斯坦诺普

（HESTER LUCY STANHOPE）

Mon. / Tue. / Wed. / Thur. / Fri. / Sat. / Sun.

Date

年

月

日

The best and sweetest things in life are things you cannot buy: the music of the birds at dawn, the rainbow in the sky. The dazzling magic of the stars, the miracle of light.

生命中最美好甜蜜的东西是买不到的：黎明时分鸟儿的啁啾，天空中的彩虹，璀璨星辰的魔力，光的奇迹。

——司屈朗

（PATIENCE STRONG）

Date

年

月

日

No land belongs unto me, yet I can go out in the meadows and see the healthy green grass – and behold the shower fall, and he that feels this, who can say he is poor?

没有一块土地是属于我的，但我却能走到草地上，凝视青葱的嫩草，看见阵雨骤下。可以感受到这些的人，谁能说他是贫穷的呢？

——约翰 · 克莱尔

（JOHN CLARE）

Mon. / Tue. / Wed. / Thur. / Fri. / Sat. / Sun.

Date

年

月

日

A persistent simplification will create an inner and outer well-being that places harmony in one's life. For me this began with the discovery of the meaninglessness of possessions beyond my actual and immediate needs...

持续简化将创造内在和外在的幸福，让你的生活充满和谐。对我而言，这开始于当我发现超出实际和眼前需求之外的财产毫无意义之时……

——和平朝圣者（PEACE PILGRIM）

Date
年
月
日

If people would stay as they really are, taking or leaving delights would make no difference. But if they will not rest in their right state, the delights develop like malignant tumours.

如果人们能够守住本心，那么得失就没有什么分别；但如果人们不能保持本心，得失便会像恶性肿瘤一样将他们吞噬。

——托马斯·默顿

（THOMAS MERTON）

Mon. / Tue. / Wed. / Thur. / Fri. / Sat. / Sun.

Date

年

月

日

Travel lightly. Happiness is inclined to get lost in the luggage.

出行可以行装简便，否则快乐容易在行李中丢失。

——帕姆·布朗

（PAM BROWN）

Date

年

月

日

The sparrow is sorry for the peacock for the burden of his tail.

麻雀看见孔雀负担着它的翎尾，替它担忧。

——拉宾德拉纳特·泰戈尔

（RABINDRANATH TAGORE）

Mon. / Tue. / Wed. / Thur. / Fri. / Sat. / Sun.

Date

年

月

日

...while we still have the song of the birds, the smell of the cut grass, and gentle refreshing breezes, life is good.

只要我们还能听到小鸟的歌唱，还能闻到草坪修剪后的味道，还能感受到温柔拂面的清风，生活就是美好的。

——摘自《弗朗西斯·盖的友情之书》

（FROM “THE FRIENDSHIP BOOK OF FRANCIS GAY”）

Date

年

月

日

I take great pleasure from the simple things – like the glint of a river and the flash of a kingfisher passing by.

我从简单的事物中获得了很大的快乐，比如说波光粼粼的小河和翠鸟飞过留下的闪亮。

——艾伦·蒂施马奇

（ALAN TITCHMARSH）

Mon. / Tue. / Wed. / Thur. / Fri. / Sat. / Sun.

Date

年

月

日

I leave this notice on my door for each accustomed visitor: "I am gone into the fields to take what this sweet hour yields; reflection, you may come tomorrow..."

我要在门前留个字条，对每个经常的来客写道："我已经到田野去漫步，享受这一刻带来的幸福。'沉思'呵，你可在明日来访……"

——珀西·比希·雪莱

（PERCY BYSSHE SHELLEY）

Mon. / Tue. / Wed. / Thur. / Fri. / Sat. / Sun.

Date

年

月

日

The hills not only take us away from a complex mode of existence, but they teach us that to be happy it is only necessary to have food, shelter and warmth. They bring us face to face with realities, and in doing so inculcate a valuable lesson in the association of simplicity and happiness.

爬山不仅可以带我们远离复杂的生活方式，还可以让我们明白只要有简单的温饱就足以快乐生活。它让我们直面现实，告诉我们一个弥足珍贵的道理：简单即幸福。

——弗兰克·悉尼·斯迈思

（FRANK. S. SMYTHE）

Mon. / Tue. / Wed. / Thur. / Fri. / Sat. / Sun.

Date

年

月

日

My crown is called content, a crown that seldom kings enjoy.

我的王冠名为知足，少有国王能够得到。

——威廉·莎士比亚

（WILLIAM SHAKESPEARE）

Date

年

月

日

Remember this – that very little is needed to make a happy life.

要记住，营造幸福生活并不需要很多东西。

——马可·奥勒留

（MARCUS AURELIUS）

Mon. / Tue. / Wed. / Thur. / Fri. / Sat. / Sun.

Date

年

月

日

To have enough is happiness, to have more than enough is harmful.

祸莫大于不知足，咎莫大于欲得。

——老子

（LAO TZU）

Date

年

月

日

Who is wise? They that learn from everyone. Who is powerful? They that govern their passions. Who is rich? They that are content.

哪种人是明智的？能向他人学习的人；哪种人是强大的？能掌控自己情感的人；哪种人是富有的？知足的人。

——本杰明·富兰克林

（BENJAMIN FRANKLIN）

Mon. / Tue. / Wed. / Thur. / Fri. / Sat. / Sun.

Date

年

月

日

Have nothing in your houses that you do not know to be useful, or believe to be beautiful.

不要在家里放任何你不知道有什么用，或是不觉得好看的东西。

——威廉·莫里斯

（WILLIAM MORRIS）

Date

年

月

日

All the truly deep people have at the core of their being the genius to be simple or to know how to seek simplicity.

真正深刻的人心怀大智慧，他们知道要简单做人，或者说知道怎样追求简单。

——马丁·马蒂

（MARTIN MARTY）

Mon. / Tue. / Wed. / Thur. / Fri. / Sat. / Sun.

Date

年

月

日

And while life was often hard and resources scarce, we always knew who we were and that the measure of our worth was inside our heads and hearts and not outside in our possessions or on our backs.

尽管生活艰难、资源不足，我们要一直明白自己是谁，明白我们的价值在于我们的头脑和心灵，而不在于我们的财富与华服。

——玛丽安·赖特·埃德尔曼

（MARIAN WRIGHT EDELMAN）

Date

年

月

日

The service of the fruit is precious, the service of the flower is sweet, but let my service be the service of the leaves in its shade of humble devotion.

果实的事业是珍贵的，花朵的事业是甜美的，但是让我做叶的事业吧，叶是谦逊的，专心地垂着绿荫。

——拉宾德拉纳特·泰戈尔

（RABINDRANATH TAGORE）

Mon. / Tue. / Wed. / Thur. / Fri. / Sat. / Sun.

Date

年

月

日

Think contentment the greatest wealth.

要把知足当作最宝贵的财富。

——乔治·谢利

（GEORGE SHELLEY）

Date

年

月

日

Crossing a bare common, in snow puddles, at twilight, under a clouded sky, without having in my thoughts any occurrence of special good fortune, I have enjoyed a perfect exhilaration. I am glad to the brink of fear.

我穿过空旷的公地，踩过积雪的水坑。时值薄暮时分，天空云层密布，虽然心中此时并没有什么鸿运当头的想法，却感受到极度的喜悦，甚至到了生出畏惧的边缘。

——拉尔夫·沃尔多·爱默生

（RALPH WALDO EMERSON）

Mon. / Tue. / Wed. / Thur. / Fri. / Sat. / Sun.

Date

年

月

日

The cool stream flowed on half asleep and quiescent in beauty, old cottages pondered benignly their musk and blown roses, the elms launched their great shadow ships floating deep green and silver. Over meadows sun-golden.

清凉的溪水在美景中半睡半醒地流淌。古老的村舍似在温柔地思忖着周遭的麝香与盛开的玫瑰。榆树用它那巨船似的树影，在铺满金色阳光的草地上，漂浮上一抹深绿和银光。

——克劳德·科利尔·阿博特

（CLAUDE COLLEER ABBOTT）

Date

年

月

日

There must be more to life than having everything!

生命中一定要有比拥有一切更重要的事。

——莫里斯·桑达克

（MAURICE SENDAK）

Mon. / Tue. / Wed. / Thur. / Fri. / Sat. / Sun.

Date

年

月

日

The single flower has more inherent beauty than the bouquet.

一朵花比一束花更具内在美。

——智慧

（WISDOM）

Date

年

月

日

Take care that something in your house shall be always wanting, whereof the deprivation is not too painful, while to wish for it is a pleasure. We must keep ourselves in such a condition that we can never be either satisfied or insatiable.

要注意让你家中某些东西时刻处于不足的状态，没有了并不会太过痛苦，期待则能带来乐趣。我们要让自己保持一种既不会满足也不会永远得不到满足的状态。

——约瑟夫·儒贝尔（JOSEPH JOUBERT）

Mon. / Tue. / Wed. / Thur. / Fri. / Sat. / Sun.

Date

年

月

日

Keep close to earth, in that lies strength. Simplicity of heart is just as necessary to an architect as for a farmer or a minister if the architect is going to build great buildings.

要贴近大地，大地中蕴含着能量。如果一位建筑师希望设计出伟大的建筑，那他一定要有一颗简单的心灵，这对农民或大臣来说都是如此。

——安·赖特（ANN WRIGHT）

Date

年

月

日

...but it is important that you are not manipulated by the makers and image creators whose job it is to make you want ever more, ever newer possessions and get caught in the trap of believing that your happiness resides in the future and your sense of meaning will be increased by the next possession you acquire.

重要的是，你不要被那些制造商和形象设计者操控。他们的工作就是让你渴望得到更多、更新的东西，然后掉入他们的陷阱，让你相信你的幸福会在将来实现，你的人生意义将因为你下一件购买的商品得到升华。

——肯特·内本（KENT NERBURN）

Mon. / Tue. / Wed. / Thur. / Fri. / Sat. / Sun.

Date

年

月

日

Simplicity rests in our rejection of greed – our ability to hold lightly all we have and let it go without resentment.

简单源自对贪欲的抛弃，也就是能看淡我们所拥有的东西，在放下的时候也毫无怨恨。

——珍妮·德·弗里斯（JENNY DE VRIES）

Mon. / Tue. / Wed. / Thur. / Fri. / Sat. / Sun.

Date

年

月

日

There is something wonderfully luxurious about sitting down to eat a pile of simple, natural produce, cooked lightly if at all, on a warm day with a beer or a glass of very cold white wine.

在一个温暖的日子里，坐下来吃一些简单烹调的天然食品，喝一杯啤酒或冰镇白葡萄酒，这可真是一种奢侈的享受了。

——马克斯·黑斯廷斯

（MAX HASTINGS）

Mon. / Tue. / Wed. / Thur. / Fri. / Sat. / Sun.

Date

年

月

日

Our minds are like crows. They pick up everything that glitters, no matter how uncomfortable our nests get with all that metal in them.

我们的思想就像乌鸦，会叼取所有闪光的东西，无论这些硬邦邦的东西会让自己的巢穴变得多不舒服。

——托马斯·默顿

（THOMAS MERTON）

Mon. / Tue. / Wed. / Thur. / Fri. / Sat. / Sun.

Date

年

月

日

Take a day off. Put the studies, the career on hold. Hide the clock. Enjoy the sunlight. Be utterly content – just for a little while.

请一天假，暂停工作和学习，把钟表藏起来，享受阳光。感受完全的满足，哪怕只有一小会儿。

——帕姆·布朗

（PAM BROWN）

Mon. / Tue. / Wed. / Thur. / Fri. / Sat. / Sun.

Date

年

月

日

One of life's great secrets is surely to celebrate the small – the sight of a flower growing between the cracks in a pavement, the look on a young child's face when they're given a birthday present they've waited for long weeks to open. Or something as simple as immersing ourselves in our favourite piece of music.

人生最美妙的秘密，肯定有一个是为小事而开心雀跃：看到人行道砖缝间开出的小花，孩子们收到期待已久的生日礼物时面露的喜色。或者，就像沉醉于最爱的音乐中这么简单。

——摘自《弗朗西斯·盖的友情之书》

（FROM "THE FRIENDSHIP BOOK OF FRANCIS GAY"）

Date

年

月

日

Beauty and perfection change and are transformed continually. Only what is simple and natural defies change.

美丽和完美一直在改变，只有简单和自然无须改变。

——厄恩斯特 · 弗里德里希 · 舒马赫

（E.F. SCHUMACHER）

Mon. / Tue. / Wed. / Thur. / Fri. / Sat. / Sun.

Date

年

月

日

I believe that we all have a job to do. It doesn't have to be grand. It's really about bringing up a family, it's about love, it's about relationships and friendships, voting, speaking your mind when you're asked. It's small things.

我相信我们都会有事要做的，不一定非得是什么宏图伟业。撑起一个家，去付出爱，处理好亲情与友情，或是选举，抑或是在被问到时讲出自己的想法，诸如此类的小事。

——斯汀（STING）

Mon. / Tue. / Wed. / Thur. / Fri. / Sat. / Sun.

Date

年

月

日

It is the simplest things in life that hold the most wonder; the colour of the sea, the sand between your toes, the laughter of a child.

在生活中，往往最简单的东西最美妙：大海的颜色、脚趾间的沙子和孩子的笑声。

——戈尔迪·霍恩

（GOLDIE HAWN）

Mon. / Tue. / Wed. / Thur. / Fri. / Sat. / Sun.

Date

年

月

日

Would that there were an award for people who come to understand the concept of enough. Good enough. Successful enough. Thin enough. Rich enough. Socially responsible enough.

就应该有一个奖项颁给那些明白什么是足够的人，比如足够好、足够成功、足够瘦、足够富有、足够具有社会责任感。

——盖尔·希伊

（GAIL SHEEHY）

Date

年

月

日

Have much and be confused.

多则惑。

——老子

（LAO TZU）

Mon. / Tue. / Wed. / Thur. / Fri. / Sat. / Sun.

Date

年

月

日

There is great joy in little things. Starlings. The first green of spring. Rain after dryness. Sun after storm. Value them. Treasure them.

小事情中有大快乐：飞过的椋鸟，春天的第一抹绿，久旱后的第一场雨，雷电过后的阳光。要珍惜、珍重这些。

——夏洛特·格雷

（CHARLOTTE GRAY）

Mon. / Tue. / Wed. / Thur. / Fri. / Sat. / Sun.

Date

年

月

日

There is great happiness in not wanting, in not being something, in not going somewhere.

没有永无止境的欲望，不过分追求功成名就，放弃不切实际的目标，我们会更加幸福。

——吉杜·克里希那穆提

（JIDDU KRISHNAMURTI）

Mon. / Tue. / Wed. / Thur. / Fri. / Sat. / Sun.

Date

年

月

日

We live in a magical fairyland and it is given to us to enjoy as an absolute gift. You do not need money in your pocket to walk through a field of wild flowers or on a heather moor… We have more blessings than we could ever count.

我们生活在一个神奇的仙境中，这绝对是上天赐给我们的礼物。我们穿过开满野花的田野或帚石楠沼泽，都不需要付钱……我们拥有的幸福比我们能想到的要多得多。

——阿尔弗雷德·温赖特

（ALFRED WAINWRIGHT）

Date

年

月

日

Small acts of helping others, if one could, small ways of making one's own life better: acts of love, acts of tea, acts of laughter. Clever people might laugh at such simplicity, but, she asked herself, what was their own solution?

在力所能及的范围内帮助他人的小举动；使自己的生活变得更好的小方法：付出爱，请人喝茶，发出笑声。聪明的人可能会嘲笑这些简单，但是，请问，要不然又该怎样让生活变得更美好呢？

——亚历山大·麦考尔·史密斯

（ALEXANDER MCCALL SMITH）

Mon. / Tue. / Wed. / Thur. / Fri. / Sat. / Sun.

Date

年

月

日

The wise do not surround themselves with clutter...They keep a place of silence, emptiness and peace.

聪明的人不会让自己身边杂乱无章……他们会让周围保持平和、空旷且宁静。

——帕姆·布朗

（PAN BROWN）

Date

年

月

日

When we simplify our lives, we can give our time and attention to what matters most.

我们只要简化自己的生活，就能把时间和精力放到最重要的事情上去。

——艾内斯·艾斯华伦

（EKNATH EASWARAN）

Mon. / Tue. / Wed. / Thur. / Fri. / Sat. / Sun.

Date

年

月

日

We gabble words like parrots until we lose the sense of their meaning, we chase after this new idea and that; we take an old thought and dress it out in so many words that the thought itself is lost in its clothing...

我们像鹦鹉那样喋喋不休，自己却始终不明白其中的含义。我们追逐一个又一个新的点子。我们找来一个过时的想法，然后用诸多词汇来装扮它，以至于思想也迷失在了文字的外壳上……

——劳拉·英格尔斯·怀尔德

（LAURA INGALLS WILDER）

Date
年
月
日

Contentment turns all it toucheth into gold; the poor person is rich with it, the rich person poor without it.

满足可以把与之相接触的一切变成金子。穷人懂得满足就能富有，富人不懂满足就是贫穷。

——谚语
（PROVERB）

Mon. / Tue. / Wed. / Thur. / Fri. / Sat. / Sun.

Date

年

月

日

To watch the corn grow, and the blossoms set; to draw hard breath over ploughshare or spade; to read, to think, to love, to hope, to pray – these are the things that make people happy.

看玉米生长，花朵绽放；倚着犁铧或铁锹，歇一口气；阅读，沉思，爱，希望，祈祷——这些都是令人快乐的事。

——约翰·拉斯金

（JOHN RUSKIN）

Mon. / Tue. / Wed. / Thur. / Fri. / Sat. / Sun.

Date

年

月

日

Contentment, as it is a short road and pleasant, has great delight and little trouble.

知足无须费力求取，它能带给人极大的快乐和极少的麻烦。

——爱比克泰德

（EPICTETUS）

Mon. / Tue. / Wed. / Thur. / Fri. / Sat. / Sun.

Date

年

月

日

It is impossible for any thinking person to look down from a hill on to a crowded plain and not ponder over the relative importance of things. To take a simple view is to take a wider view.

让任何一个会思考的人站在山顶，向那熙熙攘攘的平原望去，他不可能不去思考哪些事情更为重要？要找到简单就要让视野更开阔。

——弗兰克·悉尼·斯迈思

（FRANK. S. SMYTHE）

Mon. / Tue. / Wed. / Thur. / Fri. / Sat. / Sun.

Date

年

月

日

That person is richest whose pleasures are the cheapest.

能在最廉价的东西中找到快乐的人才是最富有的人。

——亨利·戴维·梭罗

（HENRY DAVID THOREAU）

Mon. / Tue. / Wed. / Thur. / Fri. / Sat. / Sun.

Date

年

月

日

As we value our happiness let us not forget it, for one of the greatest lessons in life is learning to be happy without the things we cannot or should not have.

如果我们珍视快乐，那就不要忘记。因为生活的一大智慧就是，学会在得不到那些我们无法或不应得到的东西时，仍然保持快乐。

——理查德·路易斯·埃文斯

（RICHARD L. EVANS）

Date

年

月

日

If you haven't all the things that you want in this world, just be grateful for the things you don't have that you never wanted.

如果你无法拥有你想得到的所有东西，那就感恩你也不曾获得你不想要的东西。

——谚语

（PROVERB）

Mon. / Tue. / Wed. / Thur. / Fri. / Sat. / Sun.

Date

年

月

日

...you need to hear past their voice, to the quieter wisdom that says you will value your possessions more if you have fewer of them, and that you will find deeper meaning in human sharing than in accumulation of goods.

你需要摒弃周围的喧嚣，聆听更平静的慧语箴言：财物越少，你就会越珍惜；与积累财富相比，与人分享更有意义。

——肯特·内本

（KENT NERBURN）

Mon. / Tue. / Wed. / Thur. / Fri. / Sat. / Sun.

Date

年

月

日

Happiness comes of the capacity to feel deeply, to enjoy simply, to think freely, to risk life, to be needed.

幸福来自深刻感受，享受简单，自由思考，敢于冒险和被人需要。

——玛格丽特·斯道姆·詹姆森

（STORM JAMESON）

Mon. / Tue. / Wed. / Thur. / Fri. / Sat. / Sun.

Date

年

月

日

I have found such joy in things that fill my quiet days – a curtain's blowing grace, a growing plant upon a window sill, a rose fresh-cut and placed within a vase, a table cleared, a lamp beside a chair, and books I long have loved beside me there.

我在那些填满我平静日子的东西中找到了快乐：被风轻轻吹起的窗帘，窗台上生机勃勃的绿植，花瓶中刚刚摘下的玫瑰，整洁的饭桌，椅旁的台灯和那些放在手边爱读的书籍。

——格雷丝 · 诺尔 · 克罗韦尔

（GRACE NOLL CROWELL）

Date

年

月

日

White hyacinths in a blue bowl. A cold-beaded glass of water. New fallen and untrampled snow. Cirrus streaked across the sky. Linen sheets. A pear tree in the spring.

在蓝色盆中的白色风信子，玻璃杯中的冰水，不久前落下还未被踩踏过的雪，天空中飘过的卷云，亚麻的桌布，春天里的一棵梨树。

——帕姆·布朗

（PAM BROWN）

Mon. / Tue. / Wed. / Thur. / Fri. / Sat. / Sun.

Date

年

月

日

Do we really need much more than this? To honour the dawn. To visit a garden. To talk to a friend. To contemplate a cloud. To cherish a meal. To bow our heads before the mystery of the day. Are these not enough?

我们真的还需要更多吗？歌颂黄昏，游览花园，与朋友谈笑，凝视一朵云，珍惜一餐饭，赞叹每天的奥秘，这些还不够吗？

——肯特·内本

（KENT NERBURN）

Mon. / Tue. / Wed. / Thur. / Fri. / Sat. / Sun.

Date

年

月

日

Live simply so that others may simply live.

生活得简单一点，这样别人就可以简单地生活。

——圣雄甘地

（MAHATMA GANDHI）

Mon. / Tue. / Wed. / Thur. / Fri. / Sat. / Sun.

Date

年

月

日

The best things in life aren't things.

生命中最宝贵的并不是物质。

——阿尔特·巴克沃德

（ART BUCHWALD）

Date

年

月

日

It becomes necessary to learn how to clear the mind of all clouds, to free it of all useless ballast and debris by dismissing the burden of too much concern with material things.

学习如何驱散头脑中的阴云至关重要，要卸下世俗烦恼的负担，扫开沉重的巨石和杂乱的残骸。

——英德拉·黛维

（INDRA DEVI）

Mon. / Tue. / Wed. / Thur. / Fri. / Sat. / Sun.

Date

年

月

日

There are two tragedies in life. One is not to get your heart's desire. The other is to get it.

生活中有两种悲剧：一种是未能实现心中的渴望，另一种是实现了。

——萧伯纳

（GEORGE BERNARD SHAW）

Date

年

月

日

The truth waits for eyes unclouded by longing.　故常无欲，以观其妙。

——老子

（LAO TZU）

Mon. / Tue. / Wed. / Thur. / Fri. / Sat. / Sun.

Date

年

月

日

The constant winds of petty appetite dissipate the power of response.

如果平时不克制琐碎的欲望，遇见急事、难事就无法应对。

——乔治·桑

（GEORGE SAND）

Mon. / Tue. / Wed. / Thur. / Fri. / Sat. / Sun.

Date

年

月

日

Peace begins when expectation ends.

不再期望什么，就能获得平静。

——斯里·钦莫伊

（SRI CHINMOY）

Mon. / Tue. / Wed. / Thur. / Fri. / Sat. / Sun.

Date

年

月

日

Real wealth, which is inexhaustible and nothing to do with the bank balance, is a richness of heart and generosity of spirit.

真正的财富用之不竭，这与银行存款无关，是心灵的富足和精神的慷慨。

——龙尊居士

（DHARMACHARI NAGARAJA）

Mon. / Tue. / Wed. / Thur. / Fri. / Sat. / Sun.

Date

年

月

日

Even though you have ten thousand fields, you can only eat one measure of rice a day; even though your dwelling contain a thousand rooms, you can only use eight feet of space at night.

良田万顷，日食三升；大厦千间，夜眠八尺。

——智慧

（WISDOM）

Mon. / Tue. / Wed. / Thur. / Fri. / Sat. / Sun.

Date

年

月

日

Ask the large questions, but seek small answers. A flower, or the space between a branch and a rock, these are enough.

询问一些宏大的问题，寻求微小的答案。一朵花，或是枝条与石块之间的缝隙，这其中蕴含的深意就够了。

——肯特·内本

（KENT NERBURN）

Mon. / Tue. / Wed. / Thur. / Fri. / Sat. / Sun.

Date

年

月

日

Our world is unbelievably beautiful and alive. I will never tire of observing it, nor will I ever cease to be moved by a sublime landscape, the texture of a rock, and the power of the elements that formed them.

我们的世界美不胜收，充满活力。我从来不会对观察世界感到厌倦，永远会为这壮丽的景色、岩石的纹理以及构成这一切的力量所感动。

——安妮·科莱

（ANNE COLLET）

Mon. / Tue. / Wed. / Thur. / Fri. / Sat. / Sun.

Date

年

月

日

Enjoy the little things of life. There may come a time when you realize they were the big things.

享受生活中的小事情。或许有一天，你会发现这些小事情其实都是大事情。

——罗伯特·布罗

（ROBERT BRAULT）

Mon. / Tue. / Wed. / Thur. / Fri. / Sat. / Sun.

Date

年

月

日

Lean liberty is better than fat slavery.

自由的清贫生活好过富裕的奴隶生活。

——谚语

（PROVERB）

Mon. / Tue. / Wed. / Thur. / Fri. / Sat. / Sun.

Date

年

月

日

I have learned in the great University of Hard Knocks a philosophy that no one who has had an easy life ever acquires. I have learned to live each day as it comes, and not to borrow trouble by dreading tomorrow.

我在“逆境大学”里学到一个道理：生活安逸的人什么也得不到。我也学会了活在今天，不因担忧未来而自寻烦恼。

——多萝西·迪克斯

（DOROTHY DIX）

Mon. / Tue. / Wed. / Thur. / Fri. / Sat. / Sun.

Date

年

月

日

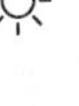

For a few days or a week or a fortnight, the fields stood "ripe to harvest." It was the one perfect period in the hamlet year... There is both beauty and bread and the seeds of bread for future generations.

再过几天、一周或是两周，就能收获田里成熟的粮食。这是一年中最美好的时光……有美景与面包，还有留给后代的种子。

——弗洛拉·汤普森（FLORA THOMPSON）

Mon. / Tue. / Wed. / Thur. / Fri. / Sat. / Sun.

Date

年

月

日

The small things that make people happy yet are too often forgotten: the colour of a bright lipstick, the scent of late-flowering sweet peas, the pleasure of a newly planted pot...

使人快乐的小事情很容易被人遗忘：口红闪亮的颜色，花期晚的豌豆的味道，刚栽下植物的花盆带来的欢乐……

——贾丝廷·皮卡迪

（JUSTINE PICARDIE）

Date

年

月

日

Those who know that enough is enough will always have enough.

知足之足，常足矣。

——老子

（LAO TZU）

Mon. / Tue. / Wed. / Thur. / Fri. / Sat. / Sun.

Date

年

月

日

I am a big man. See all these shells? They are very valuable in our culture. I could have trunks of them… But then I wouldn't be a big man. A big man gives away what he has and shares with others.

我是一个富有的人。看到这些贝壳了吗？这些在我们的文化中是非常宝贵的，而我可以拥有几大箱贝壳……但如果这么做了，我就不再是一个富有的人了。富有的人要把他拥有的东西捐出去，与他人分享。

——新几内亚岛的一位长者

（NEW GUINEA ELDER）

Date

年

月

日

Do not grasp at the stars, but do life's plain common work as it comes, certain that daily duties and daily bread are the sweetest things in life.

不要想着去摘星星，要做好生活中的每一件平常事，相信日常的工作和收获才是生活中最美好的东西。

——罗伯特·路易斯·史蒂文森

（ROBERT LOUIS STEVENSON）

Mon. / Tue. / Wed. / Thur. / Fri. / Sat. / Sun.

Date

年

月

日

Now that I am ninety-five years old, looking back over the years, I have seen many changes take place, so many inventions have been made, things now go faster, in olden times things were not so rushed. I think people were more content than they are today, you don't hear nearly as much laughter as you did in my day...

我现在 95 岁了。回顾这些年，发生了很多变化，出现了很多新的发明，事情也进展得更快了。以前，一切都没有这么匆忙。我觉得过去的人们比现在要满足得多，现在你听到的笑声可比我当年听到的少多了……

——摩西奶奶（GRANDMA MOSES）

Mon. / Tue. / Wed. / Thur. / Fri. / Sat. / Sun.

Date

年

月

日

Don't fret about getting life's grand awards. Enjoy its tiny delights.

不要为争取人生中的大奖而烦恼，享受生活中的小乐趣。

——佚名

（AUTHOR UNKNOWN）

Mon. / Tue. / Wed. / Thur. / Fri. / Sat. / Sun.

Date

年

月

日

We can't seem to see that our possessions are really butterflies that turn into caterpillars... When we finally get them, they give us a moment of elation; then, like an echo, a feeling of hollowness comes over us. The thrill of ownership grows cold in our hands.

我们似乎并没有意识到我们的财富其实是从蝴蝶变回了毛毛虫……当我们得到时，它给了我们片刻的快乐。然后，一种空虚的感觉像回声一样笼罩着我们，将财富握在手中的激动感逐渐冷却。

——肯特·内本（KENT NERBURN）

Mon. / Tue. / Wed. / Thur. / Fri. / Sat. / Sun.

Date

年

月

日

We must know how to disengage what is essential from the detail in which it is enwrapped. For everything cannot be equally considered; in a word, we must be able to simplify our duties, our business and our life.

我们要知道如何将核心内容从纠结缠绕的细节中抓取出来，因为我们并不能充分考虑到每件事。总之，我们要能够简化自己的责任、工作和生活。

——亨利·弗雷德里克·阿米埃尔
（HENRI FRÉDÉRIC AMIEL）

Mon. / Tue. / Wed. / Thur. / Fri. / Sat. / Sun.

Date

年

月

日

Possessions, outward success, publicity, luxury – to me these have always been contemptible. I believe that a simple and unassuming manner of life is best for everyone, best both for the body and the mind...

财富，外在的成功，公众的关注，奢侈品，这些对我来说都是无足轻重的。我认为简单平实的生活方式是每个人最好的选择，无论是对身体，还是对心灵而言……

——阿尔伯特·爱因斯坦

（ALBERT EINSTEIN）

My secret joy is found late at night when a million stars are reflected in the still surface of the lake. I paddle out, slip down to lie on my back in the bottom of the canoe, and drift on the water in the silence, held by a million points of light above and below.

我有个秘密的乐趣，在夜深人静时，去看漫天繁星倒映在平静的湖面上，划一艘独木舟出去，躺在舟中，静静地漂浮在水中，面前和身后被无数点星光包围。

——做梦的人

（ORIAH MOUNTAIN DREAMER）

Mon. / Tue. / Wed. / Thur. / Fri. / Sat. / Sun.

Date

年

月

日

The air is clean and bright here, cold against the skin, lifting the hair. It sweeps away the dust in my mind.

清新明净的空气，凉凉地贴在皮肤上，让人精神一振，扫去了我心中的尘埃。

——夏洛特·格雷

（CHARLOTTE GRAY）

Date
年
月
日

Blessed is the one who can enjoy the small things, the common beauties, the little day-to-day events; sunshine on the fields, birds on the bough, breakfast, dinner, supper... a friend passing by. So many people who go afield for enjoyment leave it behind them at home.

幸运的人能够享受小事情，平凡的美和日常琐事，比如田野里的阳光、树枝上的鸟儿、早餐、中餐、晚餐……一个路过的朋友等。许多到户外去寻找乐趣的人，其实把快乐留在了家里。

——大卫·加里森（DAVID GRAYSON）

Mon. / Tue. / Wed. / Thur. / Fri. / Sat. / Sun.

Date

年

月

日

The simplification of life is one of the steps to inner peace.

简化生活是走向内心平静的步骤之一。

——和平朝圣者

（PEACE PILGRIM）

Date

年

月

日

To watch the patterns move criss-cross, cast from the branches of the pear; For symphony the song of birds, and daffodils beside my chair – These are the simple joys of life; With rippling clothes put out to dry, reflecting cleanliness and light beneath a wide and changing sky.

那纵横交错的图案，是梨树枝条间投下的阴影；鸟儿乐团的交响乐，和我椅边的水仙花，这些都是生活中的小乐趣。湿漉漉的衣服，在广阔且变幻的天空下，反射着阳光，满是洁净。

——西奥多·罗斯科（THEODORA ROSCOE）

Mon. / Tue. / Wed. / Thur. / Fri. / Sat. / Sun.

Date

年

月

日

There is a master key to success with which no one can fail. Its name is simplicity... reducing to the simplest possible terms every problem.

有一把通向成功的万能钥匙，只要拥有它，就不会失败。这把钥匙就是简单……它把每个问题都简化成最简单的话语。

——亨利·德特丁

（HENRI DETERDING）

Mon. / Tue. / Wed. / Thur. / Fri. / Sat. / Sun.

Date

年

月

日

A simple kiss. The touch of a hand. A child's first smile, first step, first word. A letter. A ticket. And the world is altered.

一个简单的吻，一下触碰。孩子的第一次微笑，迈出的第一步，说出的第一个单词。一封信，一张车票，这些都会改变世界。

——帕姆·布朗

（PAM BROWN）

Mon. / Tue. / Wed. / Thur. / Fri. / Sat. / Sun.

Date

年

月

日

The incredible gift of the ordinary! Glory comes streaming from the table of daily life.

平凡是最好的礼物！荣耀来自日常生活的餐桌。

——马克丽娜·维德克尔（MACRINA WIEDERKEHR）

Mon. / Tue. / Wed. / Thur. / Fri. / Sat. / Sun.

Date

年

月

日

Those who are firm, enduring, simple and unpretentious are the nearest to virtue.

刚、毅、木、讷，近仁。

——孔子
（CONFUCIUS）

Mon. / Tue. / Wed. / Thur. / Fri. / Sat. / Sun.

Date

年

月

日

The great doing of little things makes the great life.

做好每一件小事，创造美好的生活。

——尤金妮亚·普赖斯

（EUGENIA PRICE）

Mon. / Tue. / Wed. / Thur. / Fri. / Sat. / Sun.

Date

年

月

日

When every blessed thing you have is made of silver, or of gold, you long for simple pewter.

当你拥有的一切都是金银制品时，你就会渴望简单的锡制品。

——威廉·施文克·吉尔伯特

（W. S. GILBERT）

Mon. / Tue. / Wed. / Thur. / Fri. / Sat. / Sun.

Date

年

月

日

Nothing is better than simplicity...

没有什么比简单更美好……

——沃尔特·惠特曼

（WALT WHITMAN）

Date
年
月
日

Never forget that the essence of abundance is not just material wealth. Having a grateful heart and an appreciation for life itself – for what we have made of our lives, as well as what we have been given – is the most blessed of all forms of abundance.

永远不要忘记富足的内涵不只包括物质财富。拥有一颗感恩的心，带着对生活的欣赏，也就是对我们从生活中获得的感悟以及我们所得到的一切保持感恩，这才是所有形式的富足中最能带来幸福感的。

——克里斯蒂亚娜·诺思拉普

（CHRISTIANE NORTHRUP）

Mon. / Tue. / Wed. / Thur. / Fri. / Sat. / Sun.

Date

年

月

日

Poor and content is rich, and rich enough.

人能安贫即是富有，且已足够富有。

——威廉·莎士比亚

（WILLIAM SHAKESPEARE）

Mon. / Tue. / Wed. / Thur. / Fri. / Sat. / Sun.

Date

年

月

日

What we do during our working hours determines what we have; what we do in our leisure hours determines what we are.

我们工作时间做的事，决定我们拥有什么；我们闲暇时间做的事，决定我们成为哪种人。

——乔治·伊斯曼

（GEORGE EASTMAN）

Mon. / Tue. / Wed. / Thur. / Fri. / Sat. / Sun.

Date

年

月

日

My dear girls, I am ambitious for you, but not to have you make a dash in the world – marry rich men merely because they are rich, or have splendid houses, which are not homes because love is wanting.

我亲爱的姑娘们，我对你们期望很高，但不是要你们追名逐利，比如仅仅因为钱财，或者漂亮的房子，就嫁给有钱人，这样的房子并不是家，因为里面没有爱。

——路易莎·梅·奥尔科特
（LOUISA MAY ALCOTT）

Mon. / Tue. / Wed. / Thur. / Fri. / Sat. / Sun.

Date

年

月

日

Simplicity and naturalness are the truest marks of distinction.

简单和自然是卓越最真实的标志。

——威廉・萨默塞特・毛姆

（W. SOMERSET MAUGHAM）

Mon. / Tue. / Wed. / Thur. / Fri. / Sat. / Sun.

Date

年

月

日

It is thought great to be born in palaces, surrounded with wealth – but to be born in nature's domain is greater still! I would much more glory in this birthplace, with the broad canopy of heaven above me, and the giant arms of the forest trees for my shelter, than to be born in palaces of marble, studded with pillars of gold!

人们认为出生在满是财宝的宫殿里非常幸福，但我却觉得出生在大自然里更为美好！比起镶嵌着金柱的大理石宫殿，我更为这个上有辽阔苍穹，下有大树巨大臂膀为我遮阴的出生地而感到自豪！

——乔治·考培威

（GEORGE COPWAY）

Mon. / Tue. / Wed. / Thur. / Fri. / Sat. / Sun.

Date

年

月

日

He knew how to be poor without the least hint of squalor or inelegance... He chose to be rich by making his wants few.

他知道怎样在贫穷的生活中保持清净正直……他选择通过减少自己的欲望来实现心灵的富足。

——拉尔夫·沃尔多·爱默生

（RALPH WALDO EMERSON）

Mon. / Tue. / Wed. / Thur. / Fri. / Sat. / Sun.

Date

年

月

日

We depend upon wealth and glory – the so-called good things for joy. They are illusive. The joy you receive from them is a flicker. Next second the light is gone and you are enveloped in the darkness…

我们依靠一些所谓的好东西来获得快乐，比如财富和荣耀，但这都是虚幻的。你从中得到的快乐像一闪而过的光，下一秒，光消失了，你就会被黑暗笼罩……

——斯瓦米·拉姆达斯

（PAPA RAMDAS）

Mon. / Tue. / Wed. / Thur. / Fri. / Sat. / Sun.

Date

年

月

日

Peace is something tangible. It silences the outgoing energy of the mind…

平静是实实在在的东西，能安抚想要溜走的精神能量……

——斯里·钦莫伊

（SRI CHINMOY）

Mon. / Tue. / Wed. / Thur. / Fri. / Sat. / Sun.

Date

年

月

日

To live simply is to live fully – without unnecessary clutter and distraction.

活得简单就是活得充实——抛弃不必要的喧嚣和干扰。

——帕姆·布朗

（PAM BROWN）

Mon. / Tue. / Wed. / Thur. / Fri. / Sat. / Sun.

Date

年

月

日

I hope you find joy in the great things of life – but also in the little things. A flower, a song, a butterfly on your hand.

我希望你既能在生活的大事中找到快乐，也能在小事中找到欢喜，比如一朵花、一首歌和手上的一只蝴蝶。

——埃伦·列文

（ELLEN LEVINE）

Mon. / Tue. / Wed. / Thur. / Fri. / Sat. / Sun.

Date

年

月

日

Contentment comes as the infallible result of great acceptances, great humilities – of not trying to make ourselves this or that but of surrendering ourselves to the fullness of life – of letting life flow through us.

满足来自大度的接纳和谦逊。不必试图让自己这样或那样，而是放任自己享受生活的充实，让生活从我们身边缓缓流过。

——大卫·加里森

（DAVID GRAYSON）

Date

年

月

日

The real things of life that are the common possession of us all are of the greatest value; worth far more than motor cars or radio outfits; more than lands or money; and our whole store of these wonderful riches may be revealed to us by such a common, beautiful thing as a wild flower.

生活中我们共同拥有的东西才是最有价值的，远胜汽车、音响设备、土地或金钱。所有这些美妙的财富都可以通过普通而美丽的事物展现给我们，譬如一朵野花。

——劳拉·英格尔斯·怀尔德

（LAURA INGALLS WILDER）

Mon. / Tue. / Wed. / Thur. / Fri. / Sat. / Sun.

Date

年

月

日

The goal of life is living in agreement with nature.

生活的目标是与自然和谐相处。

——基提翁的芝诺

（ZENO Of CITIUM）

Mon. / Tue. / Wed. / Thur. / Fri. / Sat. / Sun.

Date

年

月

日

Money is for food and clothes and comfort, and a visit to the pictures. Money is to make happy the lives of children. Money is for security, and for dreams, and for hopes, and for purposes.

钱可以用来买食物、衣服、电影票和让人舒适的物品。钱是让孩子们幸福生活的东西。钱是用来维护安全，完成梦想，追逐希望和实现目标的东西。

——艾伦・斯图尔特・佩顿

（ALAN STEWART PATON）

Mon. / Tue. / Wed. / Thur. / Fri. / Sat. / Sun.

Date

年

月

日

Do what you can, with what you have, where you are.

在你所处的位置，用你拥有的东西，做力所能及的事。

——西奥多·罗斯福

（THEODORE ROOSEVELT）

Date

年

月

日

Finish every day and be done with it. You have done what you could; some blunders and absurdities no doubt crept in; forget them as soon as you can.

做好每一天的事，当日事当日毕。你已经尽力了，但肯定会有一些错误和蠢事发生，就尽快忘掉它们吧。

——拉尔夫·沃尔多·爱默生

（RALPH WALDO EMERSON）

Mon. / Tue. / Wed. / Thur. / Fri. / Sat. / Sun.

Date

年

月

日

Those who have little have more room in their lives for joy.

那些不大富有的人，反而有更多空间留给生活中的乐趣。

——帕姆·布朗

（PAM BROWN）

Mon. / Tue. / Wed. / Thur. / Fri. / Sat. / Sun.

Date

年

月

日

It does not matter whether one paints a picture, writes a poem, or carves a statue, simplicity is the mark of a master-hand. Don't run away with the idea that it is easy to cook simply. It requires a long apprenticeship.

一个人是否会画一幅画，写一首诗，或者刻一座雕像，都不重要，简单才是大师的标志。不要想当然地认为烹调很容易，这需要长期的练习。

——艾尔西·德·沃尔夫
（ELISE DE WOLFE）

It's the scent of the roses that fills the air, and the whispering wind blowing through my hair. It's the sparkling dew drops on the ground, and the gurgling stream that makes hardly a sound...

空气中弥漫着玫瑰的芬芳，微风轻拂过我的头发。地上的露珠晶莹剔透，潺潺的溪水静得让人听不到声音……

——伊丽莎白·安妮·德·格雷
（ELIZABETH ANNE DE GREY）

Date

年

月

日

Those moments shared with friends and loved ones, the simple things that make people smile – they're with me all the time. They're priceless...

与朋友、爱人共度的时光，让人们会心一笑的小东西——它们一直陪伴着我，是我的无价之宝。

——埃伦·麦克阿瑟

（ELLEN MACARTHUR）

Mon. / Tue. / Wed. / Thur. / Fri. / Sat. / Sun.

Date

年

月

日

To be interested in the changing seasons is a happier state of mind than to be hopelessly in love with spring.

比起只能无望地热爱春天，对变换的四季抱有兴趣是一种更加愉悦的精神状态。

——乔治·桑塔亚那

（GEORGE SANTAYANA）

Mon. / Tue. / Wed. / Thur. / Fri. / Sat. / Sun.

Date

年

月

日

The Future is something which everyone reaches at the rate of sixty minutes an hour, whatever they do, whoever they are.

无论你从事什么工作，是何身份，你向未来行进的节奏都是每小时六十分钟。

——克利夫·斯特普尔斯·刘易斯

（C. S. LEWIS）

Mon. / Tue. / Wed. / Thur. / Fri. / Sat. / Sun.

Date

年

月

日

Ill-gotten riches are to me justas clouds passing in the sky.

不义而富且贵，于我如浮云。

——孔子

（CONFUCIUS）

Mon. / Tue. / Wed. / Thur. / Fri. / Sat. / Sun.

Date

年

月

日

Better bread with water than cake with trouble.

面包配白水，也比糟心事配蛋糕滋味好。

——俄罗斯谚语

（RUSSIAN PROVERB）

Mon. / Tue. / Wed. / Thur. / Fri. / Sat. / Sun.

Date

年

月

日

I have learned to have very modest goals for society and myself; things like clean air, green grass, children with bright eyes, not being pushed around, useful work that suits one's abilities, plain tasty food, and occasional satisfying nookie.

我学会了对社会和自己都不要抱有太高的期望。清新的空气，嫩绿的小草，眼眸明亮的孩童，不用受人摆布，做力所能及的工作，普通而又美味的食物，遇到合拍的朋友，都让人心情愉快。

——保罗·古德曼（PAUL GOODMAN）

Date

年

月

日

I think that a person who is attached to riches, who lives with the worry of riches, is actually very poor.

我觉得爱慕钱财，因富有而烦恼不断的有钱人，其实十分贫穷。

——特蕾莎修女

（MOTHER TERESA）

Mon. / Tue. / Wed. / Thur. / Fri. / Sat. / Sun.

Date

年

月

日

When I did not have cattle, I did not sleep. Now that I own cattle, I cannot sleep.

家里没有牛的时候，我睡不够；现在有了牛，我睡不着了。

——茨瓦纳谚语

（SETSWANA PROVERB）

Date

年

月

日

Any development that tends to bring people into close touch with the natural order of things is of value to mankind, in as much as it helps them to gain in a sense of proportion.

任何将人类引向事物自然秩序的进步都价值非凡，这些进步让人们学会恰当行事。

——弗兰克·悉尼·斯迈思

（FRANK. S. SMYTHE）

Mon. / Tue. / Wed. / Thur. / Fri. / Sat. / Sun.

Date

年

月

日

Freedom from desire leads to inward peace.　　夫唯不争，故无尤。

——老子

（LAOTZU）

Mon. / Tue. / Wed. / Thur. / Fri. / Sat. / Sun.

Date

年

月

日

The control of emotion by self-denial tends to mature and perfect our human sensibility.

克制欲望，控制情绪，使我们的人格臻善臻美。

——托马斯·默顿

（THOMAS MERTON）

Mon. / Tue. / Wed. / Thur. / Fri. / Sat. / Sun.

Date

年

月

日

Don’t hurry, don’t worry. You’re only here for a short visit. So be sure to stop and smell the flowers.

别行色匆匆，别愁眉苦脸。人生苦短，务必稍作停歇，嗅一嗅花香。

——华特 · 赫根

（WALTER HAGEN）

Mon. / Tue. / Wed. / Thur. / Fri. / Sat. / Sun.

Date

年

月

日

Those who live simply have empty hands – always ready to give help to others.

生活简单的人总是两手空空，因为他们随时准备着向他人伸出援手。

——帕姆·布朗
（PAM BROWN）

Mon. / Tue. / Wed. / Thur. / Fri. / Sat. / Sun.

Date

年

月

日

Simplicity, simplicity, simplicity! I say, let your affairs be as two or three, and not a hundred or a thousand. ...Simplify, simplify.

简单点！简单点！简单点！我说，只专注于两三件事就够了，别同时想着百件、千件。……简化，再简化。

——亨利·戴维·梭罗

（HENRY DAVID THOREAU）

Date

年

月

日

Eat when you're hungry. Drink when you're thirsty. Sleep when you're tired.

饿则进食，渴则饮水，困则入睡。

——佛教名言

（BUDDHIST SAYING）

Mon. / Tue. / Wed. / Thur. / Fri. / Sat. / Sun.

Date

年

月

日

If you wish for happiness, do not multiply possessions, but reduce wants. It is not the one who has too little, but the one who craves more, that is poor...

想要快乐，就不要计算财富的多少，而要减少自己的欲望。真正贫穷的人，不是拥有得太少，而是渴望得太多……

——年轻的塞内卡

（SENECA THE YOUNGER）

Mon. / Tue. / Wed. / Thur. / Fri. / Sat. / Sun.

Date

年

月

日

Beauty is the expulsion of superfluities.

去除多余就是美。

——米开朗基罗

（MICHELANGELO）

I'd stopped noticing just how beautiful a shiny new conker can be, just how wonderful is the smell of damp woodland, and just what fun it is to hear the wind whistling through the trees overhead, and watch the leaves whirling down all around you. It was a real joy to be reminded of such simple delights.

我停下脚步感受，油亮的七叶树多么美丽，森林的潮湿气息多么美妙，疾风掠过头顶树梢，身旁落叶旋转飘落。留意到这些简单的乐趣，也不失为一种快乐。

——颂歌

（CAROL）

Mon. / Tue. / Wed. / Thur. / Fri. / Sat. / Sun.

Date

年

月

日

How to be happy: keep your heart free from hate, your mind from worry, live simply, expect little, give much, sing often, pray always, forget self, think of others and their feelings, fill your heart with love, scatter sunshine. These are tried links in the golden chain of contentment.

如何变得快乐：不要憎恶他人，不要焦躁担忧，简单生活，放低期待，乐善好施，常常歌唱，坚持祷告，舍己为人，将心比心，传播温暖。事实证明，这些都是知足常乐的黄金法则。

——诺曼·文森特·皮尔

（NORMAN VINCENT PEALE）

Mon. / Tue. / Wed. / Thur. / Fri. / Sat. / Sun.

Date

年

月

日

True happinessis of a retired nature, and an enemy to pomp and noise.

真正的幸福不显山露水，它摒弃浮华和喧嚣。

——约瑟夫·艾迪生

（JOSEPH ADDISON）

Date

年

月

日

They are wise who do not grieve for the things which they do not have, but rejoice for those which they have. Learn to wish that everything should come to pass exactly as it does.

智慧之人对已有事物满足，对未拥有的事物也不懊悔。人生得失有定，应淡然处之。

——爱比克泰德

（EPICTETUS）

Mon. / Tue. / Wed. / Thur. / Fri. / Sat. / Sun.

Date

年

月

日

…sometimes the stars shine more brightly seen from the gutter than from the hilltop.

有时在阴沟里看到的星星，比山顶上看到的更加闪耀。

——阿梅莉亚·洛厄尔

（AMELIA LOWELL）

Date

年

月

日

...we have a tendency to obscure the forest of simple joys with the trees of problems.

我们往往因几株烦恼之树忽略了一整片乐趣之林。

——克里斯蒂亚娜·科朗热

（CHRISTIANE COLLANGE）

Mon. / Tue. / Wed. / Thur. / Fri. / Sat. / Sun.

Date

年

月

日

There is nothing like the first ripe apple from one's tree – tight skin, crisp flesh, and spurt of sweetness.

没有任何东西像人生之树上第一颗成熟的苹果那样令人难忘，它滋味最美——外皮紧实、果肉生脆、香甜可口。

——帕姆·布朗

（PAM BROWN）

Date

年

月

日

I have known some of the most serene, sweet and poetic hours sitting on the bench in the dark, underneath an apple tree in my garden, conversing with night. Sometimes I'm bundled up, sometimes I'm barefoot, but always I am content.

坐在花园苹果树下的长凳上，与黑夜密语交谈。有时裹着毯子，有时光着脚，内心满足安宁。这是我经历过的最平静、甜蜜和浪漫的时刻。

——萨拉·班·布雷思纳克

（SARAH BAN BREATHNACH）

Mon. / Tue. / Wed. / Thur. / Fri. / Sat. / Sun.

Date

年

月

日

If you can spend a perfectly useless afternoon in a perfectly useless manner, you have learned how to live.

要享受悠闲的生活只要一种艺术家的性情，在一种全然悠闲的情绪中，去消遣一个闲暇无事的下午。

——林语堂

（LIN YUTANG）

Date

年

月

日

Go and sit in a field on a summer's day – and just look at a daisy. If you're not filled with wonder at the power that created that, you never will find it.

坐在夏日的田野里，凝视一朵雏菊。只有惊叹于造物主的神奇力量，你才能对自然之美有所察觉。

——马库斯·亚当斯

（MARCUS ADAMS）

Mon. / Tue. / Wed. / Thur. / Fri. / Sat. / Sun.

Date

年

月

日

A truly happy person is one who can enjoy the scenery while on a detour.

真正乐观之人总能在多走了弯路时，还能有心欣赏风景。

——佚名

（AUTHOR UNKNOWN）

Mon. / Tue. / Wed. / Thur. / Fri. / Sat. / Sun.

Date

年

月

日

Close your eyes. You might try saying… something like this: "The sun is shining overhead. The sky is blue and sparkling. Nature is calm and in control of the world – and I, as nature's child, am in tune with the universe."

闭上双眼，试着对自己说……像这样的："阳光灿烂，天空蔚蓝而明媚。大自然安宁祥和，万物生长。我，大自然之子，与宇宙世界和谐共鸣。"

——戴尔·卡耐基（DALE CARNEGIE）

Mon. / Tue. / Wed. / Thur. / Fri. / Sat. / Sun.

Date

年

月

日

To feel good is of more value than a room full of expensive possessions.

家财万贯，也比不上心态平和。

——约翰·莱恩（JOHN LANE）

Mon. / Tue. / Wed. / Thur. / Fri. / Sat. / Sun.

Date

年

月

日

There is only one heroism in the world: to see the world as it is and to love it.

生活中只有一种英雄主义，那就是在认清生活真相之后依然热爱生活。

——罗曼·罗兰

（ROMAIN ROLLAND）

Mon. / Tue. / Wed. / Thur. / Fri. / Sat. / Sun.

Date

年

月

日

Embrace misfortune, for it brings you the riches of gratitude and appreciation – awareness of the miracle of your existence.

拥抱不幸吧！是不幸让你懂得感恩，懂得你的人生是奇迹。

——克里斯托弗·巴特曼

（CHRISTOPHER BATEMAN）

Mon. / Tue. / Wed. / Thur. / Fri. / Sat. / Sun.

Date

年

月

日

In reflecting habitually on the stars that may forever be too far away, we miss out on the tiny successes of each day and on the triumphs available to us at ground level.

我们总是仰望遥不可及的星光，却往往忽略每天微小的进步和触手可及的收获。

——史蒂夫·鲍基特

（STEPHEN BOWKETT）

Mon. / Tue. / Wed. / Thur. / Fri. / Sat. / Sun.

Date

年

月

日

To have lived long enough to see the sun, the dapple of leaves, star-studded skies and kindly faces – to have heard the wind, birdsong, loving voices, to have tasted clear water, fresh bread, honey – is enough to make any life worth the living.

能看见太阳、斑驳的树影、群星点缀的夜空，瞧见可亲的面孔；能听见风声、鸟语声、爱的轻语；能喝到干净的水，吃到新鲜的面包、清甜的蜂蜜，也就不枉此生了。

——帕姆·布朗（PAM BROWN）

Mon. / Tue. / Wed. / Thur. / Fri. / Sat. / Sun.

Date

年

月

日

To find the air and water exhilarating; to be refreshed by a morning walk or an evening saunter; to be thrilled by the stars at night; to be elated over a bird's nest or a wildflower in spring – these are some of the rewards of a simple life.

因空气和水心生愉悦，清晨或傍晚散步后神清气爽，对夜晚的星星惊叹不已，看见春天的鸟巢或野花都欢欣雀跃——这些都是简单生活给予我们的馈赠。

——约翰·巴勒斯（JOHN BURROUGHS）

Mon. / Tue. / Wed. / Thur. / Fri. / Sat. / Sun.

Date

年

月

日

To rejoice greatly in the good of others; to love with such generosity of heart that your love is still a dear possession in absence or unkindness – these are the gifts of fortune which money cannot buy and without which money can buy nothing.

肯定他人的品德。面对他人的无动于衷、冷眼相待，仍然心胸宽广，怀有善意。这些都是命运赠送予你的礼物，无法用金钱买到。没有这些品质，就算家财万贯，也毫无意义。

——罗伯特·路易斯·史蒂文森

（ROBERT LOUIS STEVENSON）

Mon. / Tue. / Wed. / Thur. / Fri. / Sat. / Sun.

Date

年

月

日

…fame and fortune only have true worth when mixed with other ingredients – like happiness, harmony and balance in your life.

名声和财富只有与生活的快乐、和谐、平衡相伴时，才真正拥有价值。

——雷切尔·埃尔诺

（RACHEL ELNAUGH）

Mon. / Tue. / Wed. / Thur. / Fri. / Sat. / Sun.

Date

年

月

日

To know when to stop, to know when you can get no further by your own action, this is the right beginning.

知其不可奈何而安之若命，德之至也。

——庄子

（CHUANG TZU）

Date

年

月

日

One of the greatest sounds of them all – and to me it is a sound – is utter, complete silence.

世间最动听的声音——对我来说这也算是一种声音——就是完完全全的静默。

——安德烈·哥斯特兰尼兹

（ANDRÉ KOSTELANETZ）

Mon. / Tue. / Wed. / Thur. / Fri. / Sat. / Sun.

Date

年

月

日

The foolish seek happiness in the distance, the wise grow it under their feet.

愚蠢之人去远方寻找快乐，智慧之人在脚下播种快乐。

——詹姆斯·奥本海姆

（JAMES OPPENHEIM）

Date

年

月

日

Do not linger to gather flowers to keep them, but walk on, for flowers will keep themselves blooming all your way.

别为了采摘花朵停下脚步。继续前行，你的一路都会有花香相随。

——拉宾德拉纳特·泰戈尔

（RABINDRANATH TAGORE）

Mon. / Tue. / Wed. / Thur. / Fri. / Sat. / Sun.

Date

年

月

日

We have very little, so we have nothing to be preoccupied with. The more you have, the more you are occupied, the less you give. But the less you have, the more free you are...

我们拥有的很少，也就没有什么羁绊。拥有得越多，负担越多，能给予的越少；拥有得越少，反而越加自由……

——特蕾莎修女

（MOTHER TERESA）

Date
年
月
日

The busiest, most creative soul must have at its heart a place of silence, of utter simplicity and peace.

最忙碌、最富有创造力的灵魂，在其内心深处一定拥有安静的一隅，简单又和平。

——夏洛特·格雷
（CHARLOTTE GRAY）

Mon. / Tue. / Wed. / Thur. / Fri. / Sat. / Sun.

Date

年

月

日

It's strange because we live in a world where everything is about big events; the first man on the moon or whatever, but actually the things that affect us most are things that happen to people all the time – seeing a child smile for the first time, having a baby, being married, passing an exam.

很奇怪，深受大家关注的都是大事件，例如首次登月。但对我们影响更深的，其实是大家每天都在经历的事——看见小孩初绽笑容、结婚、怀孕、通过考试等。

——罗西·斯韦尔·波普

（ROSIE SWALE POPE）

Date
年
月
日

We drown in complexities. Only those who love simplicity survive free.

复杂将我们困住。只有崇尚简单的人才得以脱身。

——帕姆·布朗
（PAM BROWN）

Mon. / Tue. / Wed. / Thur. / Fri. / Sat. / Sun.

Date

年

月

日

I don't need a diamond ring to make me happy. People say, oh, that's easy for you to say now that you can afford to buy what you want. But I don't want anything. The most valuable asset is life.

我不需要钻戒来让我开心。人们可能会说："你这么有钱，当然说得轻松了。"其实我什么也不需要。我最大的财富就是我的人生。

——华莉丝·迪里

（WARIS DIRIE）

Date

年

月

日

Eternity is in the understanding that little is more than enough.

知足方得永恒。

——罗纳德·斯图尔特·托马斯

（R. S. THOMAS）

Mon. / Tue. / Wed. / Thur. / Fri. / Sat. / Sun.

Date

年

月

日

...there is joy to be found in the simple things of life.

人生小事中处处都是乐趣。

——摘自《弗朗西斯·盖的友情之书》

（FROM "THE FRIENDSHIP BOOK OF FRANCIS GAY"）

Mon. / Tue. / Wed. / Thur. / Fri. / Sat. / Sun.

Date

年

月

日

The quiet mind is richer than a crown. Sweet are the nights in careless slumber spent... the poor estate scorns fortune's angry frown. Such sweet content, such minds, such sleep, such bliss...

内心宁静胜过加冕为王，一夜无梦最为恬适……贫困可以无视富贵的怒目。如此满足，如此宁静，如此酣眠，如此幸福……

——罗伯特·格林

（ROBERT GREENE）

Mon. / Tue. / Wed. / Thur. / Fri. / Sat. / Sun.

Date

年

月

日

The boy and girl hand in hand through a meadow; the mother washing her baby; the sweet simple things in life. We have almost lost track of them.

男孩女孩手拉手走过草地；母亲为小宝贝洗澡。这些生活中的点滴美好，我们都已然忘记。

——爱德华·史泰钦

（EDWARD STEICHEN）

Date

年

月

日

The really great of the earth are always simple. Pomp and ceremony, popes and kings, are toys for children...

这世上，真正的伟大总是那么简单。浮华与仪式，教皇与国王，都不过是孩童手中的玩具……

——埃玛·戈尔德曼

（EMMA GOLDMAN）

Mon. / Tue. / Wed. / Thur. / Fri. / Sat. / Sun.

Date

年

月

日

In spite of all wanderings, happiness is always found within a narrow compass and among objects which lie within our immediate reach.

四处漫步寻觅，才发现快乐其实近在咫尺，触手可及。

——爱德华·布威·利顿

（EDWARD BULWER LYTTON）

Mon. / Tue. / Wed. / Thur. / Fri. / Sat. / Sun.

Date

年

月

日

The greatest benefit of all lies in a simple life style; the greatest happiness of all lies in contentment.

福生于无为，而患生于多欲。知足，然后富从之。

——韩婴

（HAN YING）

Mon. / Tue. / Wed. / Thur. / Fri. / Sat. / Sun.

Date

年

月

日

Aware of the suffering caused by unmindful consumption, I vow to cultivate good health, both physical and mental, for myself, my family and my society by practicing mindful eating, drinking and consuming.

我察觉到了无意识消费所带来的痛苦。所以我发誓要培养身心的健康，为了自己、家庭和社会，我要有觉知地去饮食和消费。

——智慧

（WISDOM）

Date

年

月

日

We can make our lives so like still water that beings gather about us that they might see, it may be, their own images, and so live for a moment with a clearer, perhaps even with a fiercer life, because of our quiet.

让我们的生活如同静止的水，让我们周围存在的生命可以照见自己的样子。因为我们不挑拨，不添乱，所以他们的生活可以变得更清晰；或者又因为我们不阻止，不调和，他们的生活有可能变得更激烈。

——约翰·巴特勒·叶芝

（JOHN BUTLER YEATS）

Mon. / Tue. / Wed. / Thur. / Fri. / Sat. / Sun.

Date

年

月

日

Most people seek after what they do not possess and are then enslaved by the very things they want to acquire. Only when one has ceased to need things can you truly be your own master and really exist.

大多数人苦苦追求之物，恰恰也正是牢牢将其困住的牢笼。只有摆脱欲望，才能成为自己生活的主人并真切地活着。

——安瓦尔·萨达特

（ANWAR SADAT）

Date
年
月
日

When you come out of a thing like that "a nervous breakdown", being able to write your name gives you pride and pleasure. You see how everything can be enjoyable – the beauty of a tree, the kindness of people, buying a new lipstick, taking interest in the way you look. It's like being able to breathe again.

当你从"精神崩溃"的状态中走出来，能写出自己的名字都让你感到自豪和欣喜。每件事在你看来都是如此美好——树的婀娜，人的善意，买一支新口红，好好打理自己。这种感觉简直就像重获新生。

——莱斯利·卡侬

（LESLIE CARON）

Mon. / Tue. / Wed. / Thur. / Fri. / Sat. / Sun.

Date

年

月

日

Simple truths are a relief from grand speculations. 简单的事实让人无须去做复杂的揣测。

——沃韦纳格

（LUC DE VAUVENARGUES）

Mon. / Tue. / Wed. / Thur. / Fri. / Sat. / Sun.

Date

年

月

日

The best things in life are those nearest to you: light in your eyes, flowers at your feet, duties at your hand, the path of right just before you.

生命中最美好的，也总是近在咫尺的——眼里的光芒，脚旁的花簇，手中的责任，与面前的坦途。

——罗伯特·路易斯·史蒂文森

（ROBERT LOUIS STEVENSON）

Mon. / Tue. / Wed. / Thur. / Fri. / Sat. / Sun.

Date

年

月

日

There is beauty everywhere: in the humble daisy, in the dappling of sunlight in woodland glades, in the clouds – everywhere. There are the miracles of renewal, sunset marking the end of a day and dawn heralding the start of another…

美无处不在：不起眼的雏菊，林间空地上投射的光影，飘浮的云朵……美，俯拾皆是。每天都有新生的奇迹，有标志着一天结束的日落，也有预示着另一天开始的黎明……

——艾尔弗雷德·温赖特

（ALFRED WAINWRIGHT）

Date

年

月

日

It seems to me that one of the greatest stumbling blocks in life is this constant struggle to reach, to achieve, to acquire.

在我看来，人生中最大的障碍就是不停拼命去完成、去获得、去索求。

——吉杜·克里希那穆提

（JIDDU KRISHNAMURTI）

Mon. / Tue. / Wed. / Thur. / Fri. / Sat. / Sun.

Date

年

月

日

…possessions are chameleons that change from fantasies into responsibilities once you hold them in your hands and... they take your eye from the heavens and rivet it squarely on the earth.

财富一旦握在手里，就成了变色龙。前一秒是轻飘飘的梦幻，后一秒变身沉甸甸的重责……财富让你无心仰望星空，只顾埋头脚下。

——肯特·内本（KENT NERBURN）

Mon. / Tue. / Wed. / Thur. / Fri. / Sat. / Sun.

Date

年

月

日

Contentment is not the fulfilment of what you want, but the realisation of how much you already have.

满足不在于填补欲望，而在于珍惜拥有。

——佚名

（AUTHOR UNKNOWN）

Mon. / Tue. / Wed. / Thur. / Fri. / Sat. / Sun.

Date

年

月

日

You are back from a distant journey. There is hot water in the taps, a light that comes on to your command, familiar food, known speech, loved faces, home.

结束一段长途跋涉，回到家中。水龙头里有热水，按下开关就有光亮，熟悉的食物，亲切的声音，可亲的面孔。这就是家。

——帕梅拉·达格代尔

（PAMELA DUGDALE）

Mon. / Tue. / Wed. / Thur. / Fri. / Sat. / Sun.

Date

年

月

日

My life... I began to realize, lacks this quality of significance and therefore of beauty, because there is so little empty space. The space is scribbled on; the time has been filled.

我的生活……我突然意识到，它留白太少，没有意义，也缺乏美感。我的生活被胡乱填满，时间都被占据。

——安妮·默洛·林德伯格

（ANNE MORROW LINDBERGH）

Mon. / Tue. / Wed. / Thur. / Fri. / Sat. / Sun.

Date

年

月

日

The richest rewards are found not in accomplishments or possessions but in relationships.

最大的回报不是源于取得的成就或获得的财富，而是源于你所建立起来的情感关系。

——佚名

（AUTHOR UNKNOWN）

Date

年

月

日

We are richest who are content with the least; for content is the wealth of nature.

知足常乐才是真正的富有；因为懂得知足本身就是一笔财富。

——苏格拉底

（SOCRATES）

Mon. / Tue. / Wed. / Thur. / Fri. / Sat. / Sun.

Date

年

月

日

From what I have been told the earth has given us enough food, enough water, plants, beautiful animals, enough wonderful people. None of us ever really need be poor, if we can get control of those who are greedy. Then we can see that materially everyone is already rich.

在我看来，大地已赐予了我们足够多的食物、水、植物、美丽的动物和有趣的伙伴。控制心中的贪欲，贫困也就无从谈起，每个人都已足够富裕。

——路莎·泰什

（LUISAH TEISH）

Mon. / Tue. / Wed. / Thur. / Fri. / Sat. / Sun.

Date

年

月

日

Your wealth can be stolen, but the precious riches buried deep in your soul cannot.

身外之财可能被偷走，但灵魂的富足无人能窃取。

——蜜妮·莱普顿
（MINNIE RIPERTON）

Mon. / Tue. / Wed. / Thur. / Fri. / Sat. / Sun.

Date

年

月

日

Luxury: the lust for comfort, that stealthy thing that enters the house as a guest, and then becomes a host, and then a master.

所谓奢侈就是对舒适安逸的渴望，它悄悄潜入我们的房子，原本是位客人，渐渐反客为主，最终主宰了一切。

——卡里・纪伯伦

（KAHLIL GIBRAN）

Date

年

月

日

...to hold ever before me, even in the doing of little things, the ultimate purpose toward which I am working; to meet men and women with laughter on my lips and love in my heart; to be gentle, kind, and courteous through all the hours – this is how I desire to waste wisely my days.

我愿意把时间浪费在这些事上：专心致志、目标明确地做每一件事，哪怕只是一件小事；遇到每一个人，都面露微笑，心存善意；每一刻都保持谦逊有礼。

——托马斯·德克尔

（THOMAS DEKKER）

Mon. / Tue. / Wed. / Thur. / Fri. / Sat. / Sun.

Date

年

月

日

Contentment makes the poor rich; discontentment makes the rich poor.

贫穷的人因知足而富有，富有的人因贪婪而贫穷。

——本杰明·富兰克林

（BENJAMIN FRANKLIN）

Date

年

月

日

A gracious mind has compassion and sensitive understanding. It is without greed; rather than concentrating on what is absent or missing, it is able to celebrate and give thanks for what is present.

宽厚之人能够体谅理解他人，不贪心，不纠结于无法拥有或已失去的东西，而对已有的东西心怀感激。

——约翰 · 欧多诺休

（JOHN O' DONOHUE）

Mon. / Tue. / Wed. / Thur. / Fri. / Sat. / Sun.

Date

年

月

日

What is great wealth but a kind of prison…　泼天富贵不过是一种桎梏……

——佚名

（AUTHOR UNKNOWN）

Mon. / Tue. / Wed. / Thur. / Fri. / Sat. / Sun.

Date

年

月

日

The world is not to be put in order, the world is order. It is for us to put ourselves in unison with this order.

这个世界不需要秩序，世界本身就是秩序。我们要做的，就是融入秩序。

——亨利·米勒

（HENRY MILLER）

Mon. / Tue. / Wed. / Thur. / Fri. / Sat. / Sun.

Date

年

月

日

The old people came literally to love the soil and they sat or reclined on the ground with a feeling of being close to a mothering power. It was good for the skin to touch the earth and the old people liked to remove their moccasins and walk with bare feet on the sacred earth.

老一辈人是真真切切爱着这片土地的。每当他们坐在或是躺在地上，都能感觉到一种母亲般的呵护。双脚直接接触大地的感觉是很美妙的。老一辈人喜欢脱下软帮鞋，赤脚行走在神圣的土地上。

——路德·斯坦丁·贝尔

（LUTHER STANDING BEAR）

Mon. / Tue. / Wed. / Thur. / Fri. / Sat. / Sun.

Date

年

月

日

The wise adapt themselves to circumstances, as water moulds itself to the pitcher.

智慧之人适应环境，就如同水装进了怎样的罐子里，就变成了怎样的形状。

——谚语

（PROVERB）

Mon. / Tue. / Wed. / Thur. / Fri. / Sat. / Sun.

Date

年

月

日

To lead a simple life in reasonable comfort, with a minimum of possessions, ranks high among the arts of living. It leaves us the time, resources, and freedom of mind we need for the things that give life value: loving, helping, serving, and giving.

过一种简单的生活，享有恰当的舒适、有限的财富，这堪当生活艺术之典范。简单生活，给予我们充分的时间、资源和精神自由去做为生命带来价值的事——爱护他人、帮助他人、服务他人、馈赠他人。

——艾内斯·艾斯华伦

（EKNATH EASWARAN）

Mon. / Tue. / Wed. / Thur. / Fri. / Sat. / Sun.

Date

年

月

日

Do not seek to have everything that happens happen as you wish, but wish for everything to happen as it actually does happen, and your life will be serene.

不期待凡事如愿，懂得顺其自然。生活将会变得平和、安宁。

——爱比克泰德

（EPICTETUS）

Mon. / Tue. / Wed. / Thur. / Fri. / Sat. / Sun.

Date

年

月

日

Earth has enough for every man's need, but not every man's greed.

每个人的基本需求，大地都能满足；满足不了的，是每个人的贪欲。

——圣雄甘地

（MAHATMA GANDHI）

Mon. / Tue. / Wed. / Thur. / Fri. / Sat. / Sun.

Date

年

月

日

Happiness is a rare plant, that seldom takes root on earth: few ever enjoy it, except for a brief period... But, there is an admirable substitute for it, which all may hope to attain, as its attainment depends wholly on self – and that is, a contented spirit.

快乐正如一种稀有植物，很难在地球上扎根——很少有人真正拥有，即使能拥有，也只是短短一瞬……不过，还有一个很棒的东西能代替它，这个东西大家都渴望拥有，且能否得到全看自身——那就是知足的心态。

——玛格丽特伯爵夫人

（LADY MARGUERITE BLESSINGTON）

Mon. / Tue. / Wed. / Thur. / Fri. / Sat. / Sun.

Date

年

月

日

If you wait for tomorrow, tomorrow comes. If you don't wait for tomorrow, tomorrow comes.

如果你等待明天，明天会来；如果你不等待明天，明天也会来。

——西非曼迪卡族谚语

（WEST AFRICAN MALINKE PROVERB）

Mon. / Tue. / Wed. / Thur. / Fri. / Sat. / Sun.

Date

年

月

日

I have to live very simply, but somehow I have learned to appreciate small pleasures much more keenly.

虽然生活简朴，但不知不觉，我学会了更加珍惜生活中微不足道的快乐。

——佚名

（AUTHOR UNKNOWN）

Mon. / Tue. / Wed. / Thur. / Fri. / Sat. / Sun.

Date

年

月

日

More! More! is the cry of the mistaken soul.

误入歧途的灵魂总是在呼喊着："多一点！再多一点！"

——威廉·布莱克

（WILLIAM BLAKE）

Date

年

月

日

The little things. The click of your wife's makeup bottles and brushes... a faucet running, a gust of wind in the eucalyptus, the last rain on the window. The little things are what we remember,what we know, of family life. Of life.

妻子化妆品瓶罐和刷具碰撞的叮当声，流水的龙头，刮过桉树的劲风，窗上最后一滴雨水……这些都是日常小事，但我们关于家庭和生活的全部记忆和认知，也正是来自这些日常点滴。

——理查德·卢万（RICHARD LOUV）

Mon. / Tue. / Wed. / Thur. / Fri. / Sat. / Sun.

Date

年

月

日

To know beauty one must learn to look, to consider and to wonder.

感知美，需要我们去观察、去思考、去好奇。

——帕姆·布朗

（PAM BROWN）

Mon. / Tue. / Wed. / Thur. / Fri. / Sat. / Sun.

Date

年

月

日

Cook things so you can tell what they are. Good plain food ain't committed no crime an' don't need no disguise. Fancified cooks is the criminals.

食材只有烹饪过了才算对它们有了解。平淡无奇的食材并没有犯下什么罪过，华而不实的厨子才是罪犯。

——玛丽·拉斯韦尔（MARY LASSWELL）

Mon. / Tue. / Wed. / Thur. / Fri. / Sat. / Sun.

Date

年

月

日

…there is no more cheerful sight than a tree full of bright bonny apples.

那结满鲜亮饱满果实的苹果树，简直是最令人心情飞扬的风景了。

——巴尼·巴德斯利

（BARNEY BARDSLEY）

House made of winds. House made of fur. House made of pollens. House made of flint. House made of crystals…Bless my house made of mud, resin and pine. Bless my family made of blood, marrow and bone.

风做的房子，皮毛做的房子，花粉做的房子，火石做的房子，水晶做的房子……请保佑我泥土、树脂和松木做的房子，保佑我用鲜血、骨髓和骨头铸成的家族。

——疯马酋长

（CHIEF CRAZY HORSE）

Mon. / Tue. / Wed. / Thur. / Fri. / Sat. / Sun.

Date

年

月

日

Superfluous wealth can buy superfluities only. Money is not required to buy one necessity of the soul.

多余的财富只能买到多余的东西。心灵的必需品从不靠钱财获得。

——亨利·戴维·梭罗

（HENRY DAVID THOREAU）

Mon. / Tue. / Wed. / Thur. / Fri. / Sat. / Sun.

Date

年

月

日

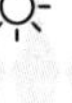

If you have to go shopping, pick up the simplest things. We have to be happy with our poverty. Let us not be driven by our small egotisms.

如果你非要买点什么的话，就买最简单的东西吧。就算贫穷，也要自得其乐。别让浅薄的自负牵着鼻子走。

——特蕾莎修女

（MOTHER TERESA）

Mon. / Tue. / Wed. / Thur. / Fri. / Sat. / Sun.

Date

年

月

日

Bloom where you are planted.

无论扎根何处，都努力绽放。

——埃拉·格拉索

（ELLA GRASSO）

Mon. / Tue. / Wed. / Thur. / Fri. / Sat. / Sun.

Date

年

月

日

Real happiness lies in gratitude. So be gratefull. Be alive. And live every moment.

真正的幸福在感恩里。所以要感恩、有活力、认真活在每一刻。

——慕尼马·马扎里（MUNIBA MAZARI)

Mon. / Tue. / Wed. / Thur. / Fri. / Sat. / Sun.

Date

年

月

日

Yes, in the poor man's garden grow far more than herbs or flowers – kind thoughts, contentment, peace of mind, and joy for weary hours.

没错，穷人的花园里不光种着草药和鲜花，还有善意、知足、安宁，和疲惫时仍然保有的快乐。

——玛丽·豪威特
（MARY HOWITT）

Mon. / Tue. / Wed. / Thur. / Fri. / Sat. / Sun.

Date

年

月

日

It is the simple things of life that make living worthwhile, the sweet fundamental things such as love and duty, work and rest and living close to nature.

生命的意义就体现在简单小事上：爱与责任，工作与休息，与大自然毗邻。正是这些最基本却也令人愉悦的事情，让生命更有价值。

——劳拉·英格斯·怀德

（LAURA INGALLS WILDER）

Mon. / Tue. / Wed. / Thur. / Fri. / Sat. / Sun.

Date

年

月

日

... I adapted well to the enforced frugal life. I found it wonderfully satisfying to be in close contact with nature every day, to feel the earth under my feet, to be away from noise and clamor.

我不得不朴素生活，但我还挺适应的。每天能和大自然亲密接触，感受脚下的土地，远离浮躁和喧嚣，感觉实在是太美妙了。

——海伦·内林（HELEN NEARING）

Date

年

月

日

Modesty and loving speech, these and none other are ornaments.

为人谦逊，言语亲和，没什么比这让人更添风采。

——提鲁瓦鲁瓦

（TIRUVALLUVAR）

Mon. / Tue. / Wed. / Thur. / Fri. / Sat. / Sun.

Date

年

月

日

A skein of geese strewn across the sky – calling. And the weary heart goes with them, flying to far-off places. Free of the earth and eager to be gone.

大雁排成一线，掠过天空，清啸回荡。疲惫的心灵也不禁随之翱翔，飞向远方，挣脱大地，找寻新的方向。

——帕姆·布朗

（PAM BROWN）

Date

年

月

日

Avoid greatness; in a cottage there may be more real happiness than kings or their favourites enjoy.

避免虚无的伟大。比起国王宫殿里的奢华享受，乡村小舍里的幸福可能更加真实。

——昆图斯·贺拉斯·弗拉库斯

（HORACE）

Mon. / Tue. / Wed. / Thur. / Fri. / Sat. / Sun.

Date

年

月

日

You are the breath of heaven and earth which goes to and fro; how can you ever possess it?

天地强阳，气也；又胡可得而有邪？

——列御寇

（LIEH TZU）

Date

年

月

日

To want more and more excitement, more possessions, we grow blind to the realities of life, to the needs of others. And when our obsessions are snatched from us – by failure, sickness or resentment – we are left with nothing but indignation, anger and a great emptiness.

为了追求更多的刺激、更多的财富，我们无视生命的真谛，无视他人的需要。一旦财富被失败、病痛或仇恨夺去，我们除了愤怒和巨大的空虚外，一无所有。

——帕梅拉·达格代尔

（PAMELA DUGDALE）

Mon. / Tue. / Wed. / Thur. / Fri. / Sat. / Sun.

Date

年

月

日

If there is to be any peace it will come through being, not having.

若宁静真的存在，那也源于顺其自然，而非靠拥有获得。

——亨利・米勒

（HENRY MILLER）

Date

年

月

日

It's the opposite of what everyone thinks. They assume that when they hang on to the things that matter in this world, they are something. But ask yourself: how could anyone who might not wake up the next morning be important?

人们以为坚持做所谓重要的事，就能成为重要人物。可事实却与此相反。你想想：连明天的太阳都未必能见到的人，还谈什么重不重要呢？

——耶切尔·米迦勒·爱泼斯坦

（RABBI YEHIEL MICHAL Of ZLOTCHOV）

Mon. / Tue. / Wed. / Thur. / Fri. / Sat. / Sun.

Date

年

月

日

A happy life is not built up of tours abroad and pleasant holidays, but of little clumps of violets noticed by the roadside.

生活的快乐并非来自出国旅行与美好的假期，路旁不经意间看到的小丛紫罗兰就足以令人心生愉悦。

——爱德华·阿德里安·威尔逊

（EDWARD A. WILSON）

Date

年

月

日

I feel gratitude every day of my life, for simple things usually. The colours of the seasons, the new life of spring... Bird-song, moonlight, quiet water, healthy children, a quiet mind, a good night's sleep. So much, so much ...

我每天都心怀感恩，多半是为了日常小事：四季的色彩、春天的生机……鸟啼、月光、水波平静、孩子健康、心境平和、一夜好眠……这样的事情，数不胜数。

——戴维·科索夫（DAVID KOSSOFF）

Mon. / Tue. / Wed. / Thur. / Fri. / Sat. / Sun.

Date

年

月

日

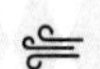

Not wanting to be anyone or anything but what we are allows us to be where we are. No longer straining our sight, not dreaming of another world, we are freed to discover this one.

决定我们处境的不是我们渴望成为谁，取得什么荣耀，而是成为真实的自己。不好高骛远，不白日做梦，便可挣脱束缚，尽情探索当下。

——迈克尔·亚当

（MICHAEL ADAM）

Date

年

月

日

People who despair because their calendars are so crowded and their duties demanding have to put a premium on simplicity. Some find a way by clearing a special room and a certain hour in which they can strip away what matters finally in their lives...

人们老是觉得生活痛苦，因为日程都排得满满当当，沉甸甸的责任让简单生活成为奢望。有人想了一招：专门空出一间屋子，腾出一定时间，让自己暂时摆脱生活中的琐碎小事……

——马丁·马蒂（MARTIN MARTY）

Mon. / Tue. / Wed. / Thur. / Fri. / Sat. / Sun.

Date

年

月

日

If you would have a successful life, less and less try to make things happen and more and more just let things happen.

要想拥有成功人生，少白费力气，多顺其自然。

——奥蒙德·麦吉尔

（ORMOND MCGILL）

Mon. / Tue. / Wed. / Thur. / Fri. / Sat. / Sun.

Date

年

月

日

Your home will have little to do with possessing or being rich.

只要有家，与拥有多少财产，以及是不是富有，并没有多大关系。

——托马斯·巴特勒姆
（THOMAS BARTRAM）

Mon. / Tue. / Wed. / Thur. / Fri. / Sat. / Sun.

Date

年

月

日

My challenge was not to do the impossible – but to learn to live with the possible.

我所面对的挑战不是去做不可能的事，而是学会接受可能的事。

——萨·本德

（SUE BENDER）

Date

年

月

日

When I am all hassled about something, I always stop and ask myself what difference it will make in the evolution of the human species in the next ten million years, and that question always helps me to get back my perspective.

每当我为某事困扰时，总会停下来想一想：在人类未来千万年的进化历程中，这点小事能产生怎样的影响呢？这么一想，我就不再庸人自扰了。

——阿内·威尔逊·施费
（ANNE WILSON SCHAEF）

Mon. / Tue. / Wed. / Thur. / Fri. / Sat. / Sun.

Date

年

月

日

Victory in a tennis match, money won in a tournament: these are not so important as good health, the honest affection and respect of friends, the love of one's wife or husband, and the spicy innocence of one's child or children.

打赢一场网球赛，获得一笔锦标赛奖金，这些都不重要，重要的是健康的体魄，朋友真诚的喜爱和尊敬，丈夫或妻子的爱以及孩子的天真烂漫。

——亚瑟·阿什（ARTHUR ASHE）

Date

年

月

日

Contentment is a pearl of great price, and whoever procures it, even at the expense of ten thousand desires, makes a wise and happy purchase.

满足感是一颗无价的珍珠。即使付出了较大的代价才得到，也是一笔明智而又能带来幸福的交易。

——约翰·巴尔戈

（JOHN BALGUY）

Mon. / Tue. / Wed. / Thur. / Fri. / Sat. / Sun.

Date

年

月

日

Slow down and enjoy life. It's not only the scenery you miss by going too fast – you also miss the sense of where you're going and why.

放慢脚步，享受生活。倘若步履匆匆，你不仅会错过一路风景，还无暇思考你去向何处，为何出发。

——埃迪·康托尔

（EDDIE CANTOR）

Date

年

月

日

The simple life is lived with an inner quietness – a deep and lasting peace.

简单生活，意味着内心的宁静——一种深沉而永恒的宁静。

——夏洛特·格雷

（CHARLOTTE GRAY）

Mon. / Tue. / Wed. / Thur. / Fri. / Sat. / Sun.

Date

年

月

日

Life is like a blanket, too short. You pull it up and your toes rebel, you yank it down and shivers meander about your shoulder; but cheerful folks manage to draw their knees up and pass a very comfortable night.

生命就像一条毯子，总嫌不够长。把它往上拉一拉，脚趾就该不乐意了；把它往下扯一扯，肩膀又冷得微微发颤。不过，乐观的家伙就会把膝盖缩起来，安安稳稳地度过一整夜。

——玛丽恩·霍华德（MARION HOWARD）

Date

年

月

日

I don't have to have millions of dollars to be happy. All I need is to have some clothes on my back, eat a decent meal when I want to, and get a little loving when I feel like it. That's the bottom line, man.

在我看来，要想幸福不一定要成为百万富翁。只要冷的时候有衣服保暖，饿的时候有东西果腹，脆弱的时候有一丝爱的温暖就够了。没错，老兄，这就是我想要的。

——雷・查尔斯（RAY CHARLES）

Mon. / Tue. / Wed. / Thur. / Fri. / Sat. / Sun.

Date

年

月

日

Sit loosely in the saddle of life.

坐在人生的马鞍上策马前行时，记得放松心态。

——罗伯特·路易斯·史蒂文森（ROBERT LOUIS STEVENSON）

Date

年

月

日

I asked for riches that I might be happy; I was given poverty that I might be wise. I asked for all things that I might enjoy life; I was given life that I might enjoy all things. I was given nothing that I asked for; But everything that I had hoped for.

我祈求财富，为我带来快乐，得到的却是贫穷，这让我收获智慧；我祈求坐拥一切，充分享受生命，得到的却只是生命本身，这让我有机会享受一切。我从未得到我祈求的一切，却收获了我渴望的所有。

——佚名（AUTHOR UNKNOWN）

Mon. / Tue. / Wed. / Thur. / Fri. / Sat. / Sun.

Date

年

月

日

We cannot get grace from gadgets.

别指望工具让我们活得优雅。

——J. B. 普瑞斯特利

（J. B. PRIESTLEY）

Date

年

月

日

I think I could turn and live with the animals, they are so placid and self contain'd, I stand and look at them long and long. They do not sweat and whine about their condition... Not one is dissatisfied, not one is demented with the mania of owning things...

我想我可以改变一下，去和动物一起生活。我站在那儿观察了它们许久，它们是那么平静而满足，不会汗流浃背地忙忙碌碌，也不会对生活怨声载道……一个个自得其乐，也没有谁因为得到了什么而欣喜若狂。

——沃尔特·惠特曼（WALT WHITMAN）

Mon. / Tue. / Wed. / Thur. / Fri. / Sat. / Sun.

Date

年

月

日

To be content with little is hard. To be content with much – impossible.

得到一点儿就满意，有些困难；拥有许多而知足，毫无可能。

——玛丽·冯·埃布讷·埃斯申巴奇

（MARIE VAN EBNER ESCHENBACH）

Mon. / Tue. / Wed. / Thur. / Fri. / Sat. / Sun.

Date

年

月

日

We collect data, things, people, ideas, "profound experiences", never penetrating any of them... But there are other times. There are times when we stop. We sit still. We lose ourselves in a pile of leaves or its memory. We listen and breezes from a whole other world begin to whisper.

我们不断积累信息、物品、人脉、想法和"深刻体验"，却从未深入感悟这一切……但某些时候，当我们停下来，静心而坐，沉浸于一堆树叶或回忆之中，屏息凝神，便能听到来自另一个世界的低声细语。

——詹姆斯·卡罗尔（JAMES CARROLL）

Mon. / Tue. / Wed. / Thur. / Fri. / Sat. / Sun.

Date

年

月

日

Having only coarse food to eat, plain water to drink, and a bent arm for a pillow, one can still find happiness therein.

饭疏食，饮水，曲肱而枕之，乐亦在其中矣。

——孔子

（CONFUCIUS）

Date

年

月

日

Money has never yet made anyone rich.

金钱从来没有让任何人富有过。

——年轻的塞内卡

（SENECA THE YOUNGER）

Mon. / Tue. / Wed. / Thur. / Fri. / Sat. / Sun.

Date

年

月

日

Abundance is not just money; it's an appreciation for life, it's sensuality. It's the love of fine things – like the beauty of a flower, a swim in a cold lake on a hot day, or a plum tree full of fruit.

生活美满不仅在于享有财富，还在于懂得欣赏生活，感知世界，并热爱那些美妙的事物——看见一朵美丽的花，炎炎夏日在清凉湖水中游上一回泳，抑或发现一棵结满果实的李子树。

——佚名（AUTHOR UNKNOWN）

Mon. / Tue. / Wed. / Thur. / Fri. / Sat. / Sun.

Date

年

月

日

Life is wonderful, and most wonderful for its smallest splendours: good coffee, children who smell like rain, bickering about whether to fix the linoleum.

生活总是充满美好，最美好的莫过于一些不起眼的小亮点——一杯香浓的咖啡，几个散发着雨的气息，为要不要修油毡而争论不休的孩子。

——杰奎琳·米恰德

（JACQUELYN MITCHARD）

Mon. / Tue. / Wed. / Thur. / Fri. / Sat. / Sun.

Date

年

月

日

Simplicity allows the heart and mind to be at peace – the senses to recognize the wonders that lie hidden all about you.

简单让身心平静——让你能够感知到内心隐藏的神奇力量。

——帕姆·布朗

（PAM BROWN）

Date

年

月

日

The thing which we speak of as beauty does not have to be sought in distant lands... It is here about us or it is nowhere...

所谓的“美”其实并不遥远……它不在别处，就在我们身边……

——艾伦·塔克

（ALLEN TUCKER）

Mon. / Tue. / Wed. / Thur. / Fri. / Sat. / Sun.

Date

年

月

日

Peace comes not from doing, but from undoing; not from getting, but from letting go...

平静不是来自埋头苦干，而是来自顺其自然；不是来自获得，而是来自舍弃……

——萨奇丹南达

（SATCHIDANANDA）

Mon. / Tue. / Wed. / Thur. / Fri. / Sat. / Sun.

Date

年

月

日

Let our advance worrying become advance thinking and planning.

把事前忧虑的时间，花在思索与计划上。

——温斯顿·丘吉尔

（WINSTON CHURCHILL）

Mon. / Tue. / Wed. / Thur. / Fri. / Sat. / Sun.

Date

年

月

日

For a greedy man even his tomb is too small. 贪婪之人连自己的坟墓都嫌小。

——塔吉克斯坦谚语

（TAJIKISTANI PROVERB）

Date

年

月

日

Manifest plainness. Embrace simplicity. Reduce selfishness. Have few desires.

见素抱朴，少私寡欲。

——老子

（LAO TZU）

Mon. / Tue. / Wed. / Thur. / Fri. / Sat. / Sun.

Date

年

月

日

Look for little pleasures, and the small, beautiful things in life. Treasure life's happy moments. Do not let them escape your notice. They are bright, golden threads in the cloth of life.

寻找生命中的那些小乐趣、小美好，珍惜生命中的每个幸福时刻，不要放过一分一秒。它们都是编织美好生活的金丝银线。

——摘自《弗朗西斯·盖的友情之书》

（FROM "THE FRIENDSHIP BOOK OF FRANCIS GAY"）

Date

年

月

日

Property is not essential. But happiness, a love of beauty, friendship between all peoples and individuals, is life itself.

财富不是必需品。幸福快乐，对美的热爱和人与人之间的真挚友谊，才是生活的本质。

——劳里·斯托克韦尔

（LAURIE STOCKWELL）

Mon. / Tue. / Wed. / Thur. / Fri. / Sat. / Sun.

Date

年

月

日

Life expresses much in the sight of a bowl of cherries, a few stems of flowers, a cup and a saucer. The gift of the every day is very dear.

一碟樱桃、几枝花、一副茶具，就足以窥见生活的意义。每个寻常日子赐予我们的礼物都弥足珍贵。

——萨拉·班·布雷思纳克

（SARAH BAN BREATHNACH）

Date

年

月

日

We build a wall about us – sealing us off from the wider world. We use sets of saucepans, mugs, teapots, ornaments and cosmetics, glass and china and wood. A multiplicity of objects. We shut ourselves off from humanity with clutter.

我们在自己周围建了一面墙，将自己隔绝于广阔世界之外。我们使用成套的炖锅、马克杯、茶壶、饰品、化妆品、玻璃、瓷器和木制品，以及各种各样的器物。这些乱七八糟的东西将我们与人性隔绝。

——帕姆·布朗（PAM BROWN）

Mon. / Tue. / Wed. / Thur. / Fri. / Sat. / Sun.

Date

年

月

日

To us, as to other spiritually-minded people in every age and race, the love of possessions is a snare, and the burdens of a complex society, a source of needless peril and temptation.

对于我们，以及其他有思想的人来说，不论年龄、不论种族，追求财富都是一个陷阱，而来自这个复杂社会的种种压力则带来了不必要的危险和诱惑。

——摘自《美国原住民的智慧》

（FROM “NATIVE AMERICAN WISDOM”）

Mon. / Tue. / Wed. / Thur. / Fri. / Sat. / Sun.

Date

年

月

日

Like water which can clearly mirror the sky and the trees only so long as its surface is undisturbed, the mind can only reflect the true image of the self when it is tranquil and wholly relaxed.

我们的思维只有在完全平静、放松的时候才能反映出真实的自己，就如同水面只有波澜不惊时，才能清晰地映射出天空和树木。

——英德拉 · 黛维（INDRA DEVI）

Date

年

月

日

I may not have a lot of money but in happiness I am a millionaire.

也许我并没有万贯家财，但论起快乐来，我就是个百万富翁。

——斯图尔特·麦克法兰和琳达·麦克法兰（STUART& LINDA MACFARLANE）

Mon. / Tue. / Wed. / Thur. / Fri. / Sat. / Sun.

Date

年

月

日

I want the same old and loved things, the same trees and soft ash green; the turtle-doves, the blackbirds, the coloured yellow-hammer sing, sing, singing so long as there is light to cast a shadow on the dial, for such is the measure of his song, and I want them in the same place.

我只想一直守着我喜爱的那些旧东西，望着老树柔和的灰绿色身影，聆听斑鸠、乌鸦和黄鹀的啁啾、轻唱，唱到日光渐稀，在钟面上留下浅浅的阴影，时间仿佛在为它们的歌声打着节拍。我希望一切就这样不变。

——理查德·杰弗里斯（RICHARD JEFFERIES）

Mon. / Tue. / Wed. / Thur. / Fri. / Sat. / Sun.

Date

年

月

日

Let your boat of life be light, packed only with what you need – a homely home and simple pleasures, one or two friends worth the name, someone to love and to love you, a cat, a dog, enough to eat and enough to wear...

生命之舟无须承载太多，只需带上你真正需要的几样东西——简单、温馨的家，一两个值得交往的好友，一个相互喜爱的人，一只小猫或小狗，还有足够的食物和衣裳……

——杰罗姆·克拉普卡·杰罗姆（JEROME K. JEROME）

Date

年

月

日

I love the look, austere, immaculate, of landscapes drawn in pearly monotones. There's something in my very blood that owns bare hills, cold silver on a sky of slate, a thread of water, churned to milky spate streaming through slanted pastures fenced with stones.

我喜欢欣赏这样的画面，以珍珠般简单的色彩绘出的朴素、无瑕的风景。我的血液里无处不流淌着对这样一番景致的向往：那里有濯濯童山，有深灰色天空露出的一丝银光，有细流涧涧，也有横跨石头围起的草坡牧场。

——埃莉诺·怀利（ELINOR WYLIE）

Mon. / Tue. / Wed. / Thur. / Fri. / Sat. / Sun.

Date

年

月

日

Simplicity is the most difficult thing to secure in this world; it is the last limit of experience and the last effort of genius.

在这个世界上，简单是最困难的事情。它既无法凭天赋获得，又无法靠经验实现。

——乔治·桑

（GEORGE SAND）

These have I loved: White plates and cups, clean-gleaming, ringed with blue lines; and feathery, faery dust; wet roofs, beneath the lamp-light; the strong crust of friendly bread; and many-tasting food; rainbows; and the blue bitter smoke of wood.

洁净发亮、点缀着蓝色线条的白色杯盘；如梦似幻、轻盈如羽的尘埃；灯光下潮湿的屋顶；嚼劲十足的面包皮；美味可口的食物；雨后的彩虹；木柴燃烧散发的苦涩的蓝色烟气——这些一直都是我喜欢的。

——鲁珀特·布鲁克（RUPERT BROOKE）

Mon. / Tue. / Wed. / Thur. / Fri. / Sat. / Sun.

Date

年

月

日

With the multiplying of your machinery you grow increasingly fatigued, anxious, nervous, dissatisfied. Whatever you have, you want more; and wherever you are you want to go somewhere else.

手边的器具多了，人反而越发疲惫、焦虑、紧张、不满。无论拥有多少，都总是渴望更多；无论身在何处，都总是向往他方。

——亚伯拉罕·米特里耶·比巴尼（ABRAHAM MITRIE RINBANY）

Date

年

月

日

To live a good and simple life one must take the variety and subtlety and astonishment of existence and translate it into one, clear, glad acceptance.

想要生活简单快乐，就请欣然接受这微妙又神奇的大千世界吧。

——帕姆·布朗

（PAM BROWN）

Mon. / Tue. / Wed. / Thur. / Fri. / Sat. / Sun.

Date

年

月

日

The great essential ingredient... is that the sacred is in the ordinary, that it is to be found in one's daily life.

最根本的一点就在于，神圣蕴藏于平凡，存在于每个人的日常生活之中。

——亚伯拉罕·哈罗德·马斯洛

（ABRAHAM HAROLD MASLOW）

Mon. / Tue. / Wed. / Thur. / Fri. / Sat. / Sun.

Date

年

月

日

There is a whole world in a single leaf.

一叶一世界。

——安迪·高兹沃斯

（ANDY GOLDSWORTHY）

Mon. / Tue. / Wed. / Thur. / Fri. / Sat. / Sun.

Date

年

月

日

Joy exists only in self acceptance. Seek perfect acceptance, not a perfect life.

唯有接纳自己，才能获得快乐。别去寻找最完美的生活，而是寻求对自我的充分肯定。

——佚名

（AUTHOR UNKNOWN）

Date
年
月
日

I know that I have worn my soul to rags and fretted my spirit to desolation, endeavoring to keep some slight track of the hurrying intellectual world about me, and I begin to know that it is a mistake.

我知道，为了努力稍稍跟上匆忙前进的世界，我已将我的灵魂折磨得疲惫不堪，任我的精神焦躁地陷入绝望。但我渐渐明白，我大错特错了。

——加马利尔·布拉德福德
（GAMALIEL BRADFORD）

Mon. / Tue. / Wed. / Thur. / Fri. / Sat. / Sun.

Date

年

月

日

Go confidently in the direction of your dreams! Live the life you've imagined. As you simplify your life, the laws of the universe will be simpler; solitude will not be solitude, poverty will not be poverty, nor weakness weakness.

自信地追逐梦想吧！去过你憧憬的生活。一旦简化生活，整个宇宙的运行法则也将变得简单。独处不再感到孤独，身无分文却不觉贫穷，脆弱也不再是弱点。

——亨利·戴维·梭罗

（HENRY DAVID THOREAU）

Mon. / Tue. / Wed. / Thur. / Fri. / Sat. / Sun.

Date

年

月

日

These are the things I prize and hold of deepest worth: light of the sapphire skies, peace of the silent hills, shelter of the forests... and the deep brown earth; but best of all, along the way, friendship and mirth.

蔚蓝明媚的天空、寂静平和的山丘、广袤隐蔽的森林……还有深深的褐色土地，都是我欣赏和珍视的东西。不过，最最重要的，还是友谊和欢笑。

——亨利·凡·戴克（HENRY VAN DYKE）

Mon. / Tue. / Wed. / Thur. / Fri. / Sat. / Sun.

Date

年

月

日

I'd rather have roses on my table than diamonds on my neck.

比起脖子上戴钻石项链，我倒更喜欢桌上摆一束玫瑰。

——埃玛·戈尔德曼（EMMA GOLDMAN）

Date

年

月

日

My earliest emotions are bound to the earth and to the labours of the fields. I find in the land a profound suggestion of poverty and I love poverty above all other things... poverty that is blessed – simple, humble, like brown bread.

我最初的情愫系于土地和田间劳作。在土地间，我领会到清贫生活的深刻教诲，在我看来，清贫远胜一切……它是一种福气——就像一块黑面包，简单又朴素。

——费德里戈·加西亚·洛尔卡
（FREDERICO GARCIA LORCA）

Mon. / Tue. / Wed. / Thur. / Fri. / Sat. / Sun.

Date

年

月

日

Of course there is no formula for success except, perhaps, an unconditional acceptance of life and what it brings.

成功当然没有公式可循。如果有，那也许就是无条件地接受生活和它带给我们的一切。

——阿图尔·鲁宾斯坦

（ARTHUR RUBINSTEIN）

Mon. / Tue. / Wed. / Thur. / Fri. / Sat. / Sun.

Date

年

月

日

Sir, respect your dinner, idolize it, enjoy it properly. You will be by many hours in the week, many weeks in the year, and many years in your life, the happier if you do.

先生，请尊重您的晚餐，崇拜它，并好好享用它。您会因此在一周中的许多时辰，一年中的许多星期，一生中的许多年都感受到更多的快乐。

——威廉·梅克比斯·萨克雷

（WILLIAM MAKEPEACE THACKERAY）

Mon. / Tue. / Wed. / Thur. / Fri. / Sat. / Sun.

Date

年

月

日

Do not indulge yourself when you are wealthy and do not mix with the vulgar world when you are in poverty. If you can draw the same pleasure from both situations, you will be free from worry and care.

不为轩冕肆志，不为穷约趋俗，其乐彼与此同，故无忧而已矣。

——庄子

（CHUANG TZU）

Mon. / Tue. / Wed. / Thur. / Fri. / Sat. / Sun.

Date

年

月

日

Voluntary simplicity involves both inner and outer conditions. It means singleness of purpose, sincerity and honesty within, as well as avoidance of exterior clutter, of many possessions irrelevant to the chief purpose of life ...

真心实意的简约需要内外兼修。它既意味着内心须从一而终、真诚坦率，也意味着须摒弃外界环境的纷繁杂乱和对人生终极目标无关紧要的身外之物。

——理查德·格雷格（RICHARD GREGG）

Mon. / Tue. / Wed. / Thur. / Fri. / Sat. / Sun.

Date

年

月

日

Teach us delight in simple things, and mirth that has no bitter springs.

学会享受简单的快乐，绽放真诚的笑容。

——鲁德亚德・吉卜林

（RUDYARD KIPLING）

Date

年

月

日

Happiness is made up of the little triumphs and small windfalls that punctuate our daily lives. It is a good dinner shared with loved ones at home; the comfort of knowing that there is a nest-egg in the bank, earned by our own honest labour...

快乐来源于一次次小小的胜利和日常生活中的意外惊喜。也许是在家和爱人共享一次美好的晚餐；也许是通过自己的辛勤劳动攒下了一笔积蓄，油然而生的欣慰之感……

——马严君玲（ADELINE YEN MAH）

Mon. / Tue. / Wed. / Thur. / Fri. / Sat. / Sun.

Date

年

月

日

I am beginning to learn that it is the sweet, simple things of life which are the real ones after all.

我渐渐开始明白，生命中真实的东西往往是美好而简单的。

——劳拉·英格斯·怀德

（LAURA INGALLS WILDER）

Date

年

月

日

My tip for the twenty-first century? Want less. Want about half what the Joneses have. If you want less stuff, you don't need as much money and then you don't need to work so hard and then you get time to have fun.

想知道我给二十一世纪人类的建议吗？那就是别渴望太多。别人想要多少，你就只要一半好了。想要的少了，需要的钱也少了，也就不必卖命工作，有时间好好享乐了。

——尼克（NICK）

Mon. / Tue. / Wed. / Thur. / Fri. / Sat. / Sun.

Date

年

月

日

I long to accomplish a great and noble task, but it is my chief duty to accomplish small tasks as if they were great and noble.

我渴望完成一件伟大而崇高的任务。但我首要的责任是把一些小任务看得同样伟大而崇高，把它们完成好。

——海伦·凯勒

（HELEN KELLER）

Mon. / Tue. / Wed. / Thur. / Fri. / Sat. / Sun.

Date

年

月

日

One day there springs up the desire for money and for all that money can provide – the superfluous, luxury in eating, luxury in dressing, trifles... The result is uncontrollable dissatisfaction.

不知从哪天起，人们开始渴望金钱和它能带来的一切——奢侈的饕餮盛宴、华丽的高档服饰和琐碎的东西……结果就是永无止境的不满足。

——特蕾莎修女（MOTHER TERESA）

Mon. / Tue. / Wed. / Thur. / Fri. / Sat. / Sun.

Date

年

月

日

It was our belief that the love of possessions is a weakness to be overcome. Its appeal is to the material part, and if allowed its way, it will in time disturb one's spiritual balance.

我们坚信拜金是一个必须克服的弱点。拜金的追求取向是物质，如果任其发展，它早晚会摧毁一个人的精神平衡。

——巴尔巴拉·米洛·尔巴赫

（BARBARA MILO OHRBACH）

Date

年

月

日

We are living on cream and meringues and luxury chocolate. Astonish yourself with good bread and clear spring water.

我们的生活正被奢侈的奶油、蛋白糖饼和巧克力包裹。去品尝一片面包，喝一口纯净水吧，你会感到惊喜的。

——帕姆·布朗（PAM BROWN）

Mon. / Tue. / Wed. / Thur. / Fri. / Sat. / Sun.

Date

年

月

日

The greatest artists and thinkers are the simplifiers.

最伟大的艺术家和思想家往往都懂得做减法。

——亨利·弗雷德里克·阿米尔

（HENRI FRÉDÉRIC AMIEL）

Date

年

月

日

Look and listen, touch and eat, smell and wander, sit and stand, pass your time in easy talk.

注视、聆听、触摸、品尝、嗅闻、游荡、坐下、起身，在轻松的交谈中任时间缓缓流逝。

——萨拉

（SARAHA）

Mon. / Tue. / Wed. / Thur. / Fri. / Sat. / Sun.

Date

年

月

日

One is happy as a result of one's own efforts, once one knows the necessary ingredients of happiness – simple tastes, a certain degree of courage, self-denial to a point, love of work, and above all, a clear conscience.

简单的餐食、一定的勇气、适当的克己忘我、对工作的热爱，以及最重要的一点，问心无愧——这些都是快乐必不可少的要素，懂得这一点的人，一定会通过努力而收获快乐。

——乔治·桑（GEORGE SAND）

Date

年

月

日

Less is more.

少即是多。

——罗伯特·勃朗宁

（ROBERT BROWNING）

Mon. / Tue. / Wed. / Thur. / Fri. / Sat. / Sun.

Date

年

月

日

To become a happy person, have a clean soul, eyes that see romance in the commonplace, a child's heart, and spiritual simplicity.

要想做个快乐的人，就要拥有一个纯净的灵魂、一双善于在平凡中发现美好的眼睛、一颗孩子般的心灵和一片简单纯粹的精神世界。

——诺曼・文森特・皮尔

（NORMAN VINCENT PEALE）

Date

年

月

日

...this was the simple happiness of complete harmony with her surroundings, the happiness that asks for nothing, that just accepts, just breathes, just is.

全身心融入周围环境的幸福就是这样，简简单单，别无他求，只是淡然接受，除了呼吸什么都不消做，就是幸福。

——伊丽莎白·亚宁

（COUNTESS VON ARNIM）

Mon. / Tue. / Wed. / Thur. / Fri. / Sat. / Sun.

Date

年

月

日

Take away the banquet. New bread. Old wine and an apple will do me very well.

别让我吃什么饕餮大餐，或者品尝什么新口味的面包。喝口老酒、啃个苹果，就最好不过了。

——帕姆·布朗

（PAM BROWN）

Date
年
月
日

... there comes a point in life when clothes, money, and a large home in the hills mean nothing, when awards aren't worth the paper they're printed on, when jewels and fancy cars are worthless ...

……到了一生之中某个时刻，你会发觉，衣服、金钱、山间豪宅都不算什么，各种奖项证书不值得印刷在纸上，珠宝和豪车也根本一文不值……

——安·玛格丽特·奥尔森
（ANN-MARGRET OLSSON）

Mon. / Tue. / Wed. / Thur. / Fri. / Sat. / Sun.

Date

年

月

日

Wealth brings the opportunity to accumulate possessions – and drive oneself mad with the managing of them.

有了钱，便可以积累财富。而管理这些财富，又会让人崩溃发狂。

——帕梅拉·达格代尔

（PAMELA DUGDALE）

Date

年

月

日

Be content with what you have; rejoice in the way things are. When you realise there is nothing lacking, the whole world belongs to you.

知足不辱，知止不殆，可以长久。

——老子

（LAO TZU）

Mon. / Tue. / Wed. / Thur. / Fri. / Sat. / Sun.

Date

年

月

日

From my father, I got the idea that wealth was not a specific figure or a set of luxurious items. That wealth was the ability to buy two ripe plums and enjoy them so fully that the memory of how good they were could last you forty years.

父亲让我懂得，财富并不是一个具体数字或是一堆奢侈品。真正的财富是，买两颗熟透的李子，大快朵颐一番，直到四十年过去了，那种幸福的味道仍记忆犹新。

——内奥米·沃尔夫（NAOMI WOLF）

Date
年
月
日

The secret of happiness is not doing what one likes to do, but in liking what one has to do.

快乐的秘诀不是做你喜欢做的，而是喜欢上你不得不做的。

——詹姆斯·马修·巴利爵士

（SIR JAMES BARRIE）

Mon. / Tue. / Wed. / Thur. / Fri. / Sat. / Sun.

Date

年

月

日

I think perhaps the most important role of education today is to combine information with lessons in moral courage... courage to risk disapproval, to change our lifestyles, to develop those qualities that lie buried in the wasteland of materialism.

我想，教育在当今最大的作用，就是在传递信息的过程中赋予人们一种内在的勇气……勇于面对质疑；勇于颠覆当前的生活状态；勇于培养那些深埋于物质主义荒原的珍贵品质。

——安德烈·科拉德（ANDRÉE COLLARD）

Date

年

月

日

…genuine happiness can be had in a plain little house, where the daily bread is earned, and some privations give sweetness to the few pleasures.

真正的幸福也许就栖身于一间朴素的小屋。在那里，只是刚刚能够维持生计，仅有的些许乐趣，却因贫穷显得甜蜜美好。

——路易莎·梅·奥尔科特（LOUISA MAY ALCOTT）

Mon. / Tue. / Wed. / Thur. / Fri. / Sat. / Sun.

Date

年

月

日

Without our familiar props, we are faced with just ourselves, a person we do not know, an unnerving stranger with whom we have been living all the time but we never really want to meet.

没有了那些赖以生存的身外之物，我们终于和真实的“我”面对面了。这个“我”是一个我们素不相识，令我们浑身不自在的陌生人，我们与之朝夕相处，却从来不愿与之谋面。

——索甲仁波切（SOGYAL RINPOCHE）

Date

年

月

日

If I were to choose the sights, the sounds, the fragrances I most would want to see and hear and smell – among all the delights of the open world – on a final day on earth, I think I would choose these: the clear, ethereal song of a white-throated sparrow singing at dawn...

假如在地球末日来临之际，要我在这大千世界中选择几样最想看到、听到和闻到的风景、声音和香味。我想我会选择：一只白颔麻雀在破晓时分天籁般清亮的吟唱……

——艾温·威·蒂尔（EDWIN WAY TEALE）

Date

年

月

日

Never get so fascinated by the extraordinary that you forget the ordinary.

非凡之物总令人着迷，但永远不要因此而遗忘了平凡小事。

——玛格达伦·纳布

（MAGDALEN NABB）

Date

年

月

日

Do you want to be happy? Then make your life as soulfully simple as sleeplessly breathing.

你想获得快乐吗？那就让生活变得极致简单吧，让每一刻就像呼吸般自然。

——斯里·钦莫伊

（SRI CHINMOY）

Mon. / Tue. / Wed. / Thur. / Fri. / Sat. / Sun.

Date

年

月

日

That which is freest, cheapest, seems somehow more valuable than anything I pay for; that which is given, better than that which is bought; that which passes between you and me in the glance of an eye, a touch of the hand, is better than minted money!

似乎免费、廉价之物总比花钱买来的有价值，获赠的就是比购买的顺眼。你我目光相接、指尖触碰的那一刻传达的，远胜过那些金银财宝！

——大卫·格雷森（DAVID GRAYSON）

Date

年

月

日

A walk in the woods with all nine of my dogs and my cocker spaniel Sophie trying to keep up; cooking fried green tomatoes with Stedman and eating them while they're hot; reading a good book and knowing another awaits.

带着我的九只狗漫步林间，还有小可卡犬索菲跟在后面，努力想要追上我们。和丈夫斯泰德曼一起煎几个青西红柿，再趁热吃掉。品读一本好书，心中清楚有另一本佳作在等着我。这些就是属于我的幸福。

——奥普拉·温弗瑞（OPRAH WINFREY）

Mon. / Tue. / Wed. / Thur. / Fri. / Sat. / Sun.

Date

年

月

日

Flow with whatever may happen and let your mind be free. Stay centered by accepting whatever you are doing. This is the ultimate.

且夫乘物以游心，托不得已以养中，至矣。

——庄子

（CHUANG TZU）

Date

年

月

日

The trouble with being in the rat race is that even if you win, you're still a rat.

和一群鼠辈拼个你死我活，即便赢了，你也还是鼠辈一个啊。

——莉莉·汤普琳

（LILY TOMLIN）

Mon. / Tue. / Wed. / Thur. / Fri. / Sat. / Sun.

Date

年

月

日

We need perfect simplicity with regard to ourselves, perfect contentment with all that comes our way, perfect peace of mind in utter self-forgetfulness.

内心深处的至简，接纳一切的知足，全然忘我的心平气和，这些都是我们需要的。

——阿尔邦·古迪尔

（ARCHBISHOP GOODIER）

Date

年

月

日

The price we pay for the complexity of life is too high...

为了应付生活的纷繁复杂，我们付出的代价实在太高了……

——让·鲍德里亚

（JEAN BAUDRILLARD）

Mon. / Tue. / Wed. / Thur. / Fri. / Sat. / Sun.

Date

年

月

日

...if I am at peace with myself, it has been a successful day.

能和自我和平相处，我就自认为成功了。

——艾利克斯·诺布尔

（ALEX NOBLE）

Date

年

月

日

Fear less, hope more; eat less, chew more; whine less, breathe more; talk less, say more; love more, and all good things will be yours.

与其恐惧，不如怀抱希望；与其贪食，不如细细品味；与其空谈，不如言之有物。常怀爱人之心，必将好事不断。

——瑞典谚语（SWEDISH PROVERB）

Mon. / Tue. / Wed. / Thur. / Fri. / Sat. / Sun.

Date

年

月

日

Money is never to be squandered or spent ostentatiously. Some of the greatest people in history have lived lives of the greatest simplicity. Remember it's the you inside that counts.

金钱从来都不是用来肆意挥霍、铺张浪费的。历史上许多伟人都过得极为简单朴素。记住，你的内心才是最重要的。

——罗丝·菲茨杰拉德·肯尼迪

（ROSE FITZGERALD KENNEDY）

Date

年

月

日

Keep a good heart. That's the most important thing in life. It's not how much money you make or what you can acquire. The art of it is to keep a good heart.

心怀善念是一生中最重要的事情。生命的意义不关乎你能挣多少钱，取得多少成就。这份善心才是生命最美好的部分。

——琼尼·米歇尔

（JONI MITCHELL）

Mon. / Tue. / Wed. / Thur. / Fri. / Sat. / Sun.

Date

年

月

日

Those who face that which is actually before them, unburdened by the past, undistracted by the future, these are they who live, who make the best use of their lives; these are those who have found the secret of contentment.

直面当下，不念过往不畏将来，才不枉活一生。这样的人才真正发挥了生命的价值，找到了知足的奥秘。

——阿尔邦·古迪尔

（ALBAN GOODIER）

Date

年

月

日

Let us follow our destiny, ebb and flow. Whatever may happen, we master fortune by accepting it.

听从命运的安排，任生命的大河潮涨潮落。无论发生什么，只有接受命运，才可掌控它。

——维吉尔

（VIRGIL）

Mon. / Tue. / Wed. / Thur. / Fri. / Sat. / Sun.

Date

年

月

日

Let life happen to you. Believe me, life is in the right, always.

顺其自然地接受你的生活吧。相信我，生活自有它的道理，不会有错的。

——赖内·马利亚·里尔克（RAINER MARIA RILKE）

Date

年

月

日

I am grateful for what I am and have. My thanksgiving is perpetual. It is surprising how contented one can be with nothing definite – only a sense of existence…

我为我的存在和所拥有的一切心存感激。对我来说，每天都是感恩节。原来只因一种虚无缥缈的东西——自我的存在感，就能如此满足。

——亨利·戴维·梭罗

（HENRY DAVID THOREAU）

Mon. / Tue. / Wed. / Thur. / Fri. / Sat. / Sun.

Date

年

月

日

When life is stripped down to its very essentials, it is surprising how simple things become. Fewer and fewer things matter and those that matter, matter a great deal more.

当生命蜕下外壳，露出本质，你会惊讶地发现一切都变得如此简单。看起来重要的事情变少了，真正重要的事情则显得愈加珍贵。

——蕾切尔·娜奥米·雷蒙

（RACHEL NAOMI REMEN）

Date

年

月

日

The little things? The little moments? They aren't little.

小事情？小时刻？它们可不容小觑。

——乔恩·卡巴特-津恩

（JON KABAT-ZINN）

365 Simpler Days: Finding a quieter, more contented life.
Published in 2014 by Helen Exley® in Great Britain
Edited by Helen Exley and illustrated by Angela Kerr.

A copy of the CIP data is available from the British Library on request.
Printed in China.
www.helenexley.com

The Simplified Chinese translation rights are arranged through RR Donnelley Asia